Positron Studies of Condensed Matter

R. N. WEST

POSITRON STUDIES OF CONDENSED MATTER

POSITRON STUDIES OF CONDENSED MATTER

R. N. WEST

School of Mathematics and Physics, University of East Anglia,
Norwich

TAYLOR & FRANCIS LTD

10–14 Macklin Street, London WC2B 5NF

1974

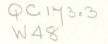
First published 1974 *by Taylor & Francis Ltd.,* 10–14 *Macklin Street, London WC2B 5NF*

Reprinted from ADVANCES IN PHYSICS, *Volume* 22, *No.* 3, *September* 1973

Printed and bound in Great Britain by Taylor & Francis Ltd., 10–14 *Macklin Street, London WC2B 5NF*

ISBN 0 85066 070 X

Distributed in the United States of America and its territories by Barnes & Noble Books (a division of Harper & Row Publishers Inc.), 10 *East* 53rd *Street, New York, N.Y.* 10022.

Preface

The first international conference on positron annihilation took place in 1965 some 35 years after the pioneering works of Anderson and Dirac. The noticeably shortening intervals that mark the arrival of the second (1971) and third (1973) of these conferences give a clear indication of the marked growth in research activity which is particularly apparent in studies of positron annihilation in condensed matter. Conference proceedings and previous reviews have furnished a wealth of reference material for the positron physicist. This monograph is aimed at a wider and less committed audience and thus attempts to isolate and analyse those particular situations in which positron studies can provide significant contributions to our understanding of condensed matter physics.

Preliminary chapters provide some theoretical background and a short account of experimental techniques. The chemical aspects of positron behaviour in some materials have been adequately dealt with in previous reviews and are considered in chapter 2 only to an extent that is necessary for completeness in the latter and larger part of this article which is concerned with studies of electronic and physical structure. Electronic structure studies have seen more than two decades of progress in theory and experiment. The discussion in chapter 3 attempts to provide a consensus if slightly conservative view of the current situation. Defected and disordered systems are a more recent and still rapidly developing area of interest. As a consequence their discussion is developed from a rather more speculative and personal viewpoint which may at some points leave room for dispute.

I would like to thank my family and various colleagues for their tolerance and good humour during the preparation of this work. I am particularly grateful to Professor W. H. Young and Dr. V. H. C. Crisp who kindly read through parts of the manuscript and supplied many helpful comments and criticisms. Thanks are also due to the many authors who so readily gave their permission for the reproduction of figures and tables. The authors are indicated in the text but it is a pleasure to acknowledge here the similarly swift response of their various publishers namely: Academic Press Inc., North Holland Publishing Co., Pergamon Press Ltd., The American Institute of Physics, The Institute of Physics, The National Research Council of Canada, and The Physical Society of Japan.

It is possible and indeed probable that at some points other authors' works have been overlooked, misinterpreted or even misrepresented. To any thus offended, I offer apologies and ask for indulgence.

October, 1973.   R. N. WEST.

Contents

PREFACE   v

INTRODUCTION   1

1. CHARACTERISTICS OF ANNIHILATION.   2
   1.1. The annihilation process.   2
   1.2. Two-photon annihilation.   5
   1.3. Experimental techniques.   8
      1.3.1. Lifetime.   8
      1.3.2. Angular correlation   10
      1.3.3. Doppler broadening.   12

2. POSITRON STATES IN MATTER.   13
   2.1. Energy considerations.   13
   2.2. Molecular materials.   15
      2.2.1. Introduction.   15
      2.2.2. Liquids   18
      2.2.3. Solids.   19
      2.2.4. Phase transitions.   23
   2.3. Ionic crystals.   25
   2.4. Metals.   27
      2.4.1. The free-electron approximation. 27
      2.4.2. Electron gas theories.   29
      2.4.3. Positron lifetimes in real metals. 38
      2.4.4. Positron effective mass and thermalization. 40

3. ELECTRONIC STRUCTURE STUDIES.   44
   3.1. The independent particle approach. 44
      3.1.1. Positron wavefunctions.   44
      3.1.2. The photon pair momentum distribution. 46
   3.2. Angular distributions and electron states in solids. 52
   3.3. Fermi surface studies.   57
      3.3.1. Background.   57
      3.3.2. Copper and its alloys.   58
      3.3.3. Simple and not so simple metals. 67
      3.3.4. Magnetized materials.   70

4. DEFECTED SOLIDS AND DISORDERED SYSTEMS. 71
   4.1. The statistical approach.   71
      4.1.1. Positron states and lifetime spectra. 71
      4.1.2. Analysis of experiments.   74
   4.2. Ionic crystals.   76
      4.2.1. Positron states in ionic materials. 76
      4.2.2. Positron traps in alkali halides. 80

4.3. Metals.   86
    4.3.1. Experiments.   86
    4.3.2. ' Theories '.   90
    4.3.3. Applications.   97
    4.3.4. Alloys.   102
    4.3.5. Liquid metals.   103
4.4. Powders, voids, surface, and other phenomena.   105

REFERENCES   111

SUBJECT INDEX

POSITRON STUDIES OF CONDENSED MATTER

Introduction

When energetic positrons enter condensed matter they rapidly lose almost all their energy by collisions with electrons and ions. After a further, somewhat longer period, characteristic of the medium, their annihilation is announced by the emergence of energetic photons whose energies, momenta, and time of emission may be measured with high precision with modern detector systems. The utility of positron annihilation studies of condensed matter relies on the fact that these characteristics of the annihilation process, which in principle involve the sophisticated considerations of quantum electrodynamics, nevertheless depend almost entirely on the initial state of the positron–many-electron system.

In some materials the positron may become bound to a particular electron to form the neutral positronium atom. The subsequent fate of the positron is then determined both by the possible states of this atom and its interactions with the surrounding medium. Some of the more complex aspects of this problem form the basis of *positronium chemistry*, a subject which has been reviewed in depth by Goldanskii (1968). In this article we shall restrict our discussions, in the main, to those modes of positronium annihilation which most directly reflect the properties of the containing medium.

More frequently we shall be concerned with the annihilation of slow quasi-free positrons for which the initial state will depend not only on the background potential system but also on the Coulomb force of the positron which sometimes may considerably perturb the original electronic system. The theoretical description of the effect of the light-charged impurity on a polarizable many-electron system is of considerable general interest and has stimulated much theoretical effort, to which this article will try to do justice.

A particular benefit gained from the many-electron theories is our increased understanding of the relevance of positron studies to the electronic structure of the unperturbed system. Positron studies of electronic structure have provided a valuable confirmation of many of the established concepts of solid-state physics and, in some cases, information not obtainable from other techniques.

A more recent development has been the full recognition of the importance of sample purity in positron studies. The remarkable sensitivity of positrons to structural defects has suggested a new range of applications for what by now are well established measurement techniques.

In subsequent chapters we shall consider these various areas of interest in some detail. To keep our account reasonably self-contained we commence with a brief account of the measurable characteristics of the annihilation process, their typical values, and how they can be related to the initial state of the positron–electron system.

§ 1. Characteristics of annihilation

1.1. *The annihilation process*

The selection rules and other considerations that determine the probability of the various modes of annihilation of slow positron–electron pairs are discussed in detail in fundamental texts on Quantum Electrodynamics (Akhiezer and Berestetskii 1965, Berestetskii *et al.* 1971) and previous reviews (Wallace 1960, Goldanskii 1968). Here we shall merely recall the essential results.

The annihilation process will supply an energy $\sim 2m_0c^2$, the total rest mass energy of the annihilating pair. For slowly moving pairs, conservation of momentum demands that at least two other bodies (particles or quanta) be involved in the process.

Fig. 1

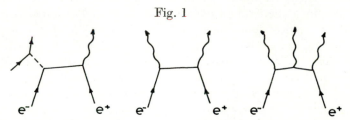

Feynmann diagrams for one, two and three-photon annihilation.

Single-photon annihilation is only possible in the presence of a third body, an electron or nucleus which can absorb the recoil momentum. Feynmann diagrams for one, two, and three-photon emission are shown in fig. 1. Since in such diagrams the introduction of an additional vertex multiplies the cross section for the process by a factor of the order of the fine structure constant, $\alpha = 1/137$, the cross section for three-photon annihilation is more than two orders of magnitude smaller than that for the two-photon process. The cross section for single-photon annihilation is still further reduced by the presence of an additional factor $\lambda_c^3\rho$, where λ_c is the Compton wavelength of the electron and ρ the density of additional atoms or electrons that can absorb the recoil momentum. The largest value of ρ likely to be encountered in any physical situation is such as to make this additional factor of order of α^3. Thus the probability of two-photon annihilation is considerably greater than that for one or three photons ; the ratios of the cross sections for the respective processes being

$$\sigma_{(3)}/\sigma_{(2)} \approx \alpha, \quad \sigma_{(1)}/\sigma_{(2)} \approx \alpha^4.$$

The cross section for the two-photon annihilation of a free positron and a stationary electron was shown by Dirac (1930) to be

$$\sigma_{(2)} = \frac{\pi r_0^2}{\gamma + 1} \left\{ \frac{\gamma^2 + 4\gamma + 1}{\gamma^2 - 1} \ln\left[\gamma + \sqrt{(\gamma^2 - 1)}\right] - \frac{\gamma + 3}{\sqrt{(\gamma^2 - 1)}} \right\}. \tag{1}$$

$\gamma = (1 - v^2/c^2)^{-1/2}$ and $r_0 = e^2/m_0 c^2$ is the classical electron radius.

At low positron energies, one obtains a cross section inversely proportional to the positron velocity v :

$$\sigma_{(2)} = \frac{\pi r_0^2 c}{v}. \tag{2}$$

Consequently, the annihilation probability

$$\Gamma_{(2)} = \sigma_{(2)} v n_e = \pi r_0^2 c n_e \tag{3}$$

is independent of the positron velocity and simply proportional to the density of electrons n_e. It should be admitted that these equations are barely applicable if the positron velocity is very low ($v/c \leqslant \alpha$), in which case the Coulomb attraction of the annihilating pair must be considered.

In eqn. (1), and those like it which result from averaging over spin directions and photon polarizations, the influence of various conservation laws is not apparent. The effect of such laws and the resulting selection rules are more easily demonstrated if we consider annihilation from well-defined initial states such as the bound states of the positronium atom (Wheeler 1946, Pirenne 1947).

The ground states of the positronium atom are the singlet ^1S state or parapositronium, in which the orbital angular momentum l and the total spin are zero, and the triplet ^3S state, orthopositronium, in which $l = 0$ and the total spin is unity. Since a system of two photons cannot have states of angular momentum equal to unity (Akhiezer and Berestetskii 1965) it follows that the triplet state cannot decay into two photons. Additional constraints are supplied by the requirements of conservation of parity.

The charge parity of the electron plus positron system is (Berestetskii *et al.* 1971) $C = P_i P_l P_s$, where P_i is the intrinsic parity which is negative for a particle–antiparticle pair, $P_l = (-1)^l$ is the spatial parity, and $P_s = (-1)^{s+1}$ is the spin parity. Thus for positronium $C = (-1)^{l+s}$ is $+1$ for the singlet state and -1 for the triplet state. The charge parity of the photon is negative and for a system of n photons $C = (-1)^n$. It follows from conservation of charge parity that parapositronium decays into an even number of photons and orthopositronium into an odd number of photons. In view of our earlier observations concerning relative cross sections it is clear that the main processes determining the lifetime of positronium are the two-photon decay of parapositronium and the three-photon decay of orthopositronium.

The ground-state positronium wavefunctions have the form

$$\psi(r) = (\pi a^3)^{-1/2} \exp\left(-r/a\right), \tag{4}$$

where r is the relative coordinate and $a = 2\hbar^2/m_0 e^2$ is the Bohr radius of positronium which is twice that for hydrogen (the reduced mass is $m_0/2$). The electron and positron momenta in this state are of order $p = a^{-1} \ll m_0 c$ which allows us to regard the system as approximating to the annihilation of free particles of zero momenta but definite spin orientation.

From eqn. (3) we deduce that the decay rate of parapositronium is

$$\Gamma_{\text{para}} = 4\pi r_0^2 c |\psi(0)|^2 \tag{5}$$

Here $|\psi(0)|^2 = (\pi a^3)^{-1}$ is the electron density at the positron and the factor four arises because eqn. (3) corresponds to the initial state averaged over spins, whereas in positronium only one of the four possible spin states can undergo two-photon decay. Substituting the values for r_0 and a we obtain a parapositronium lifetime,

$$t_{\text{para}} = \Gamma_{\text{para}}^{-1} = 123 \text{ psec.}$$

In like manner we may use the expression for the spin average cross section for three-photon annihilation (Berestetskii *et al.* 1971),

$$\sigma_{(3)} = \frac{4(\pi^2 - 9)}{3v} r_0^2 c\alpha, \tag{6}$$

to deduce a value for the orthopositronium lifetime,

$$t_{\text{ortho}} = 140 \text{ nsec.}$$

It is worth pointing out that for both the singlet and triplet states the level width, deduced from the lifetime by $\delta E = \hbar/t$, is small compared to the energy level $E = me^4/4\hbar^2$ and thus positronium may be regarded as a system in a quasi-stationary state.

A comparison of eqns. (2) and (6) reveals that the ratio of three-photon to two-photon annihilations for slow unbound pairs is $\sim 1/370$. If positronium is formed the fraction of three photon events will be larger. However, in most materials the increase will be slight since the normal decay of the orthopositronium atoms must compete with more rapid processes arising from the interactions of these orthopositronium atoms with other electrons and atoms in the system. Should, as is often the case, these interactions result in the eventual annihilation of the positron with an antiparallel spin electron from the surrounding medium useful information about the system can be obtained, as in the case of the annihilation of quasi-free positrons, by measurements of the characteristics of the two-photon mode of annihilation.

1.2. *Two-photon annihilation*

The mean lifetimes of positrons in condensed matter normally lie within the range 100 picoseconds to 10 microseconds and provide a guide to the electron density at the annihilation site. In addition to the lifetime, two other characteristics of the two-photon annihilation process, the relative emission directions and the energies of the annihilation photons, can provide useful information about the initial state of the electron–positron system.

In the centre-of-mass frame of the annihilating pair, the two photons are emitted in opposite directions, each with an energy equal to one-half the total energy of the system, i.e. $2E_0 = E_T = 2m_0c^2 -$ binding energy $\approx 2m_0c^2$. In the laboratory frame, moving at velocity $v \ll c$ (fig. 2 *b*) the energies and emission directions are changed. Application of the well-known transformations of special relativity (Rindler 1960) yields the result

$$\cot \theta_{1,\,2} = (\cos \theta_0 \pm v/c)/(1 - v^2/c^2)^{1/2} \sin \theta_0.$$

Fig. 2

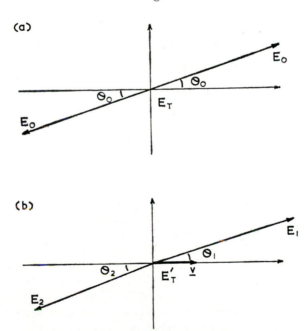

Two-photon emission. (*a*) Centre of mass frame. (*b*) Laboratory frame.

By elementary manipulation of the trigonometric functions we deduce that

$$\tan (\theta_1 - \theta_2) = \delta\theta \approx \frac{2v}{c} \sin \theta_0 + O(v^2/c^2).$$

If we ignore the small differences between the angles of emission in each frame, i.e. in $\sin \theta_0$ we set $\theta_0 = \theta_1 = \theta_2$,

$$\delta\theta \approx \frac{2v}{c} \sin \theta_1 \approx \frac{2v}{c} \sin \theta_2 \approx \frac{p_T}{m_0 c}, \tag{7}$$

where p_T is the momentum transverse to the emission direction of the annihilation photons.

The repulsive force between the ions and the charged positron will cause it to spend most of its time in the interstitial regions of a solid or liquid where it will annihilate with electrons of momenta of order $10^{-2} m_0 c$. Thus, angular correlation studies involve the measurement of deviations from antiparallelism of the order of tens of milliradians, adequate resolution of which brings attendant difficulties of small solid angles and low counting efficiency. We shall return to consider the technical details of such experiments in § 1.3.

With a similar analysis we can obtain an expression for the energies of the annihilation quanta in the laboratory frame (fig. 2 b), i.e.

$$E_{1,\,2} = \tfrac{1}{2} E_T (1 \pm v/c \cos \theta_0)/\sqrt{(1 - v^2/c^2)}$$

$$= \tfrac{1}{2} E'_T (1 \pm v/c \cos \theta_0).$$

The total energy in the laboratory frame, $E_T{}'$, contains the kinetic energy as well as the rest mass and binding energy of the pair, but again, providing $v \ll c$,

$$E_{1,\,2} = E_0 \pm \delta E \approx m_0 c^2 \pm c p_l/2. \tag{8}$$

p_l is the component of momentum 'along' the emission direction. For 10 eV electrons (and almost stationary positrons) δE is about 1·5 keV. With the advent of solid-state detectors the consequent broadening of the 0·511 MeV annihilation line can be measured with some accuracy. Although the information obtained is equivalent to that provided by angular correlation studies, the rapidity with which one can make the energy measurements points to their considerable potential in some types of defect study (see chapter 4).

In order to interpret the results of lifetime, angular correlation or Doppler broadening experiments we require an expression for the probability that a positron in an arbitrary system of electrons and external fields will annihilate with the production of two photons having total momentum **p**. The most general treatment of this problem was supplied by Chang Lee (1958) who applied a series of canonical transformations of the type suggested by Foldy and Wouthuysen (1950) to the Hamiltonian for the interaction of the particles and the radiation field. As a result, the interaction Hamiltonian splits into two parts, the first of which conserves the number of particles, and can be treated together with terms arising from the Coulomb interaction and the external field. In the second part, which deals with annihilation and creation, the effects of the external field and the Coulomb interaction enter only through the wave-functions of the system.

Since each interaction of an electron or positron with the radiation field can create only one photon the annihilation is a second-order process, and must proceed by way of an intermediate state (fig. 1). In the inter-mediate state, either the electron or the positron is recoiling with a momentum approximately equal to that of the emitted photon $p \approx m_0 c$. Consequently, the relativistic particle in the intermediate state is little affected by the external fields and may be taken as free. Further, if the initial state is extremely non-relativistic the energy denominator for the transition is essentially constant and we can sum over the intermediate states using the completeness condition. (Wallace 1960.)

It may then be shown, using the language of second quantization, that the part of the second-order Hamiltonian operator that produces a photon pair of momentum $\mathbf{p} = \hbar\mathbf{k}$, is proportional to (Ferrell 1956),

$$\sum_{\mathbf{k}_1} \sum_{\mathbf{k}_2} a_{\mathbf{k}_1} b_{\mathbf{k}_2} \delta_{\mathbf{k}_1 + \mathbf{k}_2, \mathbf{k}}.$$

$a_{\mathbf{k}_1}$ and $b_{\mathbf{k}_2}$ are the electron and positron annihilation operators in the momentum representation and the δ function arises from the photon creation operators and expresses the necessary conservation of momentum. If we insert the expressions for the corresponding position field operators $\psi_-(\mathbf{x}_1)$ and $\psi_+(\mathbf{x}_2)$,

$$a_{\mathbf{k}_1} = V^{-1/2} \int d^3\mathbf{x}_1 \exp(-i\mathbf{k}_1 . \mathbf{x}_1)\psi_-(\mathbf{x}_1),$$

$$b_{\mathbf{k}_2} = V^{-1/2} \int d^3\mathbf{x}_2 \exp(-i\mathbf{k}_2 . \mathbf{x}_2)\psi_+(\mathbf{x}_2),$$

and perform the sums over $\mathbf{k}_1$ and $\mathbf{k}_2$, we obtain the operator

$$\int d^3\mathbf{x} \exp(-i\mathbf{k} . \mathbf{x})\psi_-(\mathbf{x})\psi_+(\mathbf{x}). \tag{9}$$

According to perturbation theory, the probability per unit time of annihilation with the emission of a photon pair having 'momentum' in the range $d^3\mathbf{k}$ at $\mathbf{k}$ is then

$$\Gamma(\mathbf{k})d^3\mathbf{k} = \text{const} \left| \sum_f \langle f| \int d^3\mathbf{x} \exp(-i\mathbf{k} . \mathbf{x}) \right.$$
$$\left. \times \psi_-(\mathbf{x})\psi_+(\mathbf{x})|i\rangle \right|^2 d^3\mathbf{k}. \tag{10}$$

Here, $|i\rangle$ represents the initial state of one positron and n electrons, and

$|f\rangle$ any of the possible final states containing the remaining $(n-1)$ electrons.

Performing the sum over the final states we obtain the result

$$\Gamma(\mathbf{k})d^3\mathbf{k} = \text{const.} \int \int d^3\mathbf{x} \, d^3\mathbf{y} \, \exp\left[-i\mathbf{k} \cdot (\mathbf{x}-\mathbf{y})\right]$$

$$\times \langle \psi_-^\dagger(\mathbf{y})\psi_+^\dagger(\mathbf{y})\psi_+(\mathbf{x})\psi_-(\mathbf{x})\rangle \, d^3\mathbf{k}. \qquad (11)$$

The total annihilation rate results from the further sum over all photon momenta $\mathbf{k}$ whence,

$$\Gamma = \text{const.} \int d^3\mathbf{x}\langle \psi_-^\dagger(\mathbf{x})\psi_-(\mathbf{x})\psi_+^\dagger(\mathbf{x})\psi_+(\mathbf{x})\rangle \qquad (12)$$

Thus the annihilation rate is simply proportional to the expectation value of the electron density at the positron, averaged over all positron positions. The constant of proportionality can then be determined by comparison with eqn. (3) whence, if V is the volume of normalization,

$$\Gamma = \frac{\pi r_0^2 c}{V} \int d^3\mathbf{x}\langle \psi_-^\dagger(\mathbf{x})\psi_-(\mathbf{x})\psi_+^\dagger(\mathbf{x})\psi_+(\mathbf{x})\rangle. \qquad (13)$$

In the case that the initial state can be represented by the product of a positron wavefunction and a Slater determinant of n one-electron wavefunctions, application of the annihilation operator yields n terms (Chang Lee 1958). Each term is the product of a determinant of degree $(n-1)$ multiplied by a factor

$$\int d^3\mathbf{x}\psi_r(\mathbf{x})\psi_+(\mathbf{x}), \qquad (14)$$

where $\psi_r(\mathbf{x})$ and $\psi_+(\mathbf{x})$ are the wavefunctions of the rth electron and positron respectively. The probability of annihilation with the rth electron is then simply proportional to the square of the modulus of the Fourier transform of the wavefunction product, i.e.

$$\Gamma_r(\mathbf{k})d^3\mathbf{k} = \frac{\pi r_0^2 c}{(2\pi)^3} \left| \int d^3\mathbf{x} \exp\left(-i\mathbf{k} \cdot \mathbf{x}\right)\psi_r(\mathbf{x})\psi_+(\mathbf{x}) \right|^2 d^3\mathbf{k}. \qquad (15)$$

The total annihilation rate is obtained by summing over all occupied states and photon momenta, i.e.

$$\Gamma = \sum_r \Gamma_r = \sum_r \int \Gamma_r(\mathbf{k}) \, d^3\mathbf{k}. \qquad (16)$$

Equations (10), (11), (12) and (15) form the starting point of most calculations of the two-photon annihilation parameters in condensed matter.

1.3. Experimental techniques

1.3.1. Lifetime

Contemporary positron lifetime experiments usually make use of the isotope ^{22}Na which by the emission of a simultaneous (< 10 psec) gamma-ray provides a zero of time signal of the entry of the positron into the sample. The time delay between the detection of this $1 \cdot 28$ MeV photon and the subsequent detection of a $0 \cdot 51$ MeV annihilation photon is measured by a fast–slow coincidence technique employing time to pulse-height conversion and multichannel analysis (Bell 1966, Gedcke and McDonald 1968, Gedcke and Williams 1968).

A typical system is shown in fig. 3. The timing discriminators in the fast channels are used to process the 'raw' pulses from each detector in such a way as to define accurately the time of occurrence of each detected event. The time to pulse-height convertor produces a spectrum of pulses in which the amplitude of each pulse is proportional to the time interval between the detection of a pair of events in detectors 1 and 2. The transfer of these pulses to the multichannel analyser for sorting and storage proceeds by way of a linear gate driven by coincident events in the slow energy discriminating channels which thus provide a degree of identification or isolation of relevant pairs of events.

Fig. 3

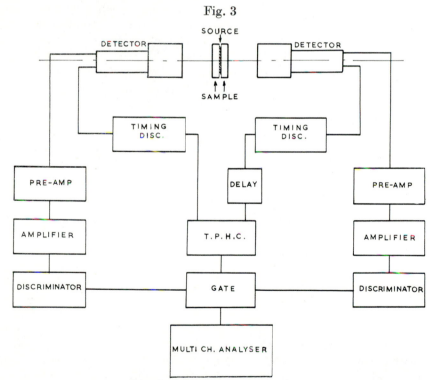

A typical lifetime spectrometer.

Ideally, the random or systematic errors in the measuring process should be much smaller than the measured delays. In practice, in positron studies, the stored spectrum of delays is significantly smeared by such errors and can be regarded as the result of a convolution of the real spec-

trum with the time resolution function of the apparatus.   A close approximation to the instrumental resolution function can be obtained by replac-
ing, without other adjustments, the positron source by a source such as
^{60}Co which emits two simultaneous gamma-rays.

The measurement of lifetimes of the order of a few hundred picoseconds
represents the limit of present-day measurement techniques and the ultimate success of experiments depends greatly on source and sample
preparation, good electronic stability and realistic statistical analysis.
The realization of a satisfactory coincidence counting rate without a
significant accidental background contribution demands a modest source
intensity and compact geometry.   In the most usual arrangement the
two detectors are placed either side of a thin sandwich composed of a few
microcuries of ^{22}Na and two layers of the sample of interest.

Commerically available ^{22}Na sources are frequently supplied as aqueous
solutions of sodium chloride made weakly acidic to prevent sodium ion
transfer to the containing vessel.   When, as is frequently the case, these
sources are evaporated directly onto the sample surface they may attack
the latter, producing contaminated surface layers in which the positrons
can annihilate and confuse an intended measurement of the lifetime in the
bulk material (Kugel *et al.* 1966).   If the source is contained between foils
of more attack-resistant materials such as gold, mica, or polyester it is
necessary that these be extremely thin (~ 1 mg cm^{-2}) so as to allow a
satisfactory transmission of positrons (Bertolaccini and Zappa 1967).
With either arrangement a necessary part of any data analysis will be the
identification of the (hopefully) small intensity contribution from annihilation in the source or source assembly.

The measurement of *relative mean* lifetimes in different materials may be
obtained from the time displacement between the centroid of the delayed
coincident curve and that of a ' prompt ' curve such as that for ^{60}Co (Bay
1950).   The reliability of this approach was limited but recently has been
much improved by the introduction of a procedure whereby the prompt
and delayed curves may be simultaneously accumulated (Knapton and
McKee 1971).

Where the lifetime spectrum is believed to consist of a limited number of
exponential decay components of the type $\exp(-\lambda t)$ a statistically
weighted curve-fitting procedure may be warranted (Kirkegaard and
Eldrup 1971, 1972 ; Lichtenberger *et al.* 1971).   It would however be
difficult to devise a sentence which could do justice to the problems that
can be encountered in such an analysis (Lanzcos 1957, Eldrup *et al.* 1971).
In cases where the individual components have similar decay rates
exceptional electronic stability and statistical accuracy are required (Crisp
et al. 1973).

1.3.2. *Angular correlation*

A typical experimental set-up for the measurement of the angular
correlation of annihilation photons is shown in fig. 4.   The radioactive

source ^{64}Cu, ^{22}Na, or ^{58}Co is placed close to the sample or, where this is impractical, the positrons are focused onto the sample by the application of a suitable magnetic field. In measurements on copper or its alloys an *in situ* ^{64}Cu source may be employed. The annihilation photons are detected by scintillation counters shielded from direct view of the source by lead collimators. Additional detector collimators define the instrumental angular resolution. The coincident (< 10 nsec) counting rate from the two detectors is measured as a function of the displacement of one detector. In order to minimize errors resulting from source decay and electronic drift an angular distribution is usually obtained from the results of many cycles of the movable detector through the necessary angular range.

Fig. 4

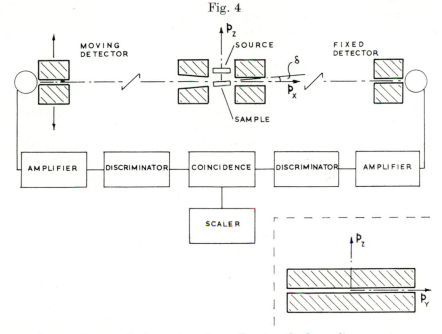

An angular correlation apparatus. Inset : the long slit geometry.

The natural coincidence between the 0·51 MeV photons allows large (10 millicurie—1 curie) sources to be employed with the retention of a good true to accidental background ratio. The most common compromise between good angular resolution and adequate counting rates is provided by the long slit geometry (fig. 4). Here, one dimension of the detector slit is made very much larger than the width of a typical angular distribution (for a discussion of this point see Mogensen 1970) and thus only one component, p_z, of the transverse momentum of the photon pair remains defined. Some alternative geometries that have been utilized for Fermi surface studies are discussed in § 3.3.2.

For a fixed sample-detector distance, the ultimate angular resolution is limited by the penetration of the positron beam into the adjacent sample

surface.   This penetration and the attenuation of the 0·51 MeV photons
in the sample will produce an assymmetric resolution curve if the sample
surface lies in the p_x, p_y plane (Mijnarends 1969).   These effects can be
largely removed by rotating the specimen through a small angle ~ 40
milliradians (fig. 4).   Such a rotation also prevents the distortion of the
angular distributions from single crystal samples which can arise from the
diffraction of the annihilation photons (Hyodo et al. 1971).

Precise measurements can involve an optical angular resolution $\sim 0·1$
milliradians and the accumulation of $\sim 10^5$ coincidence counts per position
in the neighbourhood of the peak of the angular distribution.   A sophisticated analysis of the data is then possible.

1.3.3. *Doppler broadening*

A sensible measurement of the longitudinal component of the photon
pair momentum (eqn. (8)) can only be made with a solid-state detector
(Hotz et al. 1968, Rama Reddy and Carrigan 1970).   At the time of
writing commerically available Ge(Li) detectors can provide an energy
resolution $\sim 1·8$ keV for 1·33 MeV ^{60}Co gamma-rays.   At 0·51 MeV, the
consequent resolution of the photon momentum distribution is equivalent
to that of a long slit angular correlation apparatus having an angular
resolution ~ 4 milliradians.   Despite this handicap the necessity for only
one detector results in the rapid accumulation of data and allows for the
investigation of samples in difficult environmental conditions or of a form
unsuitable for angular correlation studies.

Fig. 5

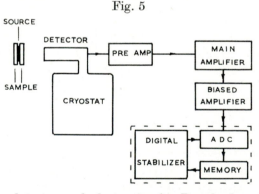

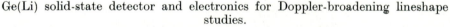

Ge(Li) solid-state detector and electronics for Doppler-broadening lineshape
studies.

The experimental arrangement shown in fig. 5 may be regarded as basic.
The biased amplifier should provide for sufficient expansion of the energy
scale to allow many analyser channels to be encompassed by the Doppler
broadened curve whose width can then be measured with great accuracy.

The very short time that this technique has been applied to positron
studies has probably not allowed its full potential to be realized.   Neither
has it allowed the possible complications and problems to be fully assessed.
Already it is clear that sophisticated stabilization techniques and careful

count-control precautions are required if subtle changes in the width of
the energy peak are to be recorded.

§ 2. POSITRON STATES IN MATTER

2.1. *Energy considerations*

A study of the behaviour of positrons in condensed matter provides a
stimulating problem in its own right.   It can also guide us to those situations in which the characteristics of the annihilation process are likely to
provide useful information about electronic and other properties of the
system.   The behaviour of the positron, prior to annihilation, is strongly
reflected in the form of the characteristic lifetime spectra (fig. 6).   For the
purpose of the present discussion, it is convenient to classify the materials
of interest according to the following scheme :

    (*a*) Molecular materials.   (*b*) Ionic crystals.   (*c*) Metals.

Fig. 6

TIME IN NSEC

Typical lifetime spectra.   (*a*) Molecular materials.   (*b*) Ionic solids.   (*c*)
    Metals.   Note the variation in time scales for these figures.

In (*a*) we include condensed gases, organic solids and liquids and glassy
or amorphous insulators.   These materials usually exhibit lifetime spectra
that can be resolved into two or three components with lifetimes ranging
from 0·1 to 10 nanoseconds.   A comprehensive table of data obtained
prior to 1967 has been compiled by Hogg and Laidlaw (1968).   Ionic
crystals such as metal halides, hydrides and oxides also show several
components (Bertolaccini *et al.* 1971).   Here, the individual lifetimes are
very similar and are resolved only with difficulty by the latest techniques.
Lifetimes in metals are normally characterized by a single component and
lie in the range 100–500 picoseconds (Weisberg and Berko 1967).   Ex-

amples of these three types of spectra are shown in fig. 6. The occurrence of multi-component spectra in types (*a*) and (*b*) materials suggests that the positron can annihilate from a variety of distinct states. The possible existence of some of these states can be argued purely from energy considerations.

The positrons emitted from the most commonly used isotopes ^{64}Cu, ^{22}Na, ^{58}Co have maximum energies of the order of 0·5 MeV. Such energetic positrons, on entering condensed matter, rapidly lose most of their energy by collisions with the electrons of the system. As long as the positron energy is significantly greater than the typical energies required to promote electrons to excited bands we may assume that the positron will lose of the order of half its energy in each collision. With the further assumption of a collision-free path of the order of a lattice spacing it is easy to show that the positron is reduced to an energy of a few electron volts in a time $t \approx 10^{-15}$ sec.

Further slowing down will be affected by the availability of final states for scattered electrons. An early calculation (Lee-Whiting 1955) of the time required for a positron in a metal to reach thermal energies ($\sim 0·025$ eV) produced a thermalization time of 3 picoseconds. Thus we may conclude that most of the positrons in metals are thermalized before annihilation. On the other hand, in insulators, where energy loss to electrons is inhibited by energy gaps, the final stages of thermalization can only proceed by excitation of lattice vibrations. De Benedetti *et al.* (1950) deduced a thermalization time of the order of 300 picoseconds for this process. It follows that the thermalization of free positrons may not be completed before annihilation in some materials. The final stages of the slowing-down process may be further complicated by the formation of a positronium atom which will moderate more slowly than a free positron by virtue of its comparatively weaker interactions with the surrounding medium.

According to Wallace (1960), positronium can be formed in molecular materials where electron exchange occurs only within individual molecules and covalent forces between molecules, if they exist, are repulsive. The energetics of positronium formation in solids are most easily discussed in terms of the ' Ore Gap ', a concept originating from the corresponding problem in gases (Ore 1949). The binding energy of the positronium atom is 6·8 eV in free space but may be smaller in matter. If the ionization potential of a molecule of the medium is V_i, positronium can be formed as long as the initial energy of the positron E is greater than $V_i - 6·8$ eV. If $E > V_i$, inelastic scattering with ionization will be more probable (Wallace 1960) and if $E > V_e$, the lowest excitation potential, this excitation will compete with positronium formation. Thus positronium formation is most probable for the range of positron energies

$$V_e > E > V_i - 6·8 \text{ eV}. \qquad (17)$$

This is the ' Ore Gap ' whose width will play a significant role in determining the positronium yield.

Within this range of energies other factors such as the excitation of vibrational modes and the formation of positron and molecule complexes may conspire to inhibit positronium formation (Goldanskii 1967). The introduction of a low concentration of impurities producing excitation levels within the Ore Gap may decrease the positronium yield. On the other hand, such impurities may also introduce ionization potentials lower than that of the pure substance with a consequent increase in positronium formation (Brandt and Feibus 1969). The application of electric fields may also affect positronium formation in either direction. The field may induce diffusion out of the Ore Gap, thus decreasing the positronium yield or, if sufficiently large, it may extend the energy distribution of thermalized positrons into the Gap. (Brandt and Feibus 1968). In spite of these many complications the Ore Gap analysis has proved of considerable value as a guide to the probability of positronium formation in a variety of materials and, in some cases, the experimentally observed positronium yield is close to that suggested by an Ore Gap analysis (Goldanskii 1967).

If a fraction f of the positrons leave the Ore Gap in positronium, $f/4$ will be in the singlet state and will subsequently annihilate with a lifetime $\sim t_{\mathrm{para}} = 123$ picoseconds. The remaining $3f/4$ will not necessarily annihilate with the orthopositronium lifetime of 140 nanoseconds. Various processes, which can occur during collisions of the positronium atom with the moleules of the sytems, may affect the fate of these positrons.

' Quenching ' of positronium refers to all processes which tend to shorten the positronium lifetime and includes ortho $\rightarrow$ para conversion, two-photon annihilation resulting from chemical reactions of positronium, and two-photon annihilation, on collision, with a ' foreign ' electron from the surrounding medium (pick-off). The observation of the shortened positron lifetime then provides a measure of the corresponding quenching rate.

Ortho $\rightarrow$ para conversion can result from electron exchange in a collision with a molecule having unpaired electrons (Ferrell 1958). Quenching, observed in the presence of paramagnetic molecules, may sometimes be ascribed to such conversion (Lee and Celitans 1965). More frequently quenching will result from more complex chemical interactions of the positronium atom with the medium molecules. The chemical factors affecting both the inhibition and quenching of positronium have been discussed in depth by Goldanskii 1967, 1968) and we shall be content to provide but a few simple examples in § 2.2.

Annihilation by pick-off during collisions has most relevance to the present study in so far as the annihilation occurs with an electron of the surrounding medium rather than with an electron bound to a positronium atom or complex.

2.2. *Molecular materials*

2.2.1. *Introduction*

We are now in a position to present the simplest possible picture of the origins of the various components of the lifetime spectra of molecular materials (fig. 6 a).   The longest-lived component of intensity I_3 can arise from pick-off of positrons in orthopositronium and if so, the lifetime τ_3 will be largely determined by the characteristic rate for this process.   As long as τ_3 is significantly smaller than the lifetime of orthopositronium the fraction of positrons forming positronium is $\sim \frac{4}{3}I_3$.   In consequence there should also be present a short lifetime component of intensity $\sim I_3/3$ which arises from the normal decay of parapositronium and fast pick-off from that state.   Such a component can be barely resolved with the present state of experimental measurement which involves resolving times of the order of $200 \rightarrow 300$ picoseconds (Gedcke and Williams 1968). However, when the positronium atoms have typically thermal energies a measure of the parapositronium fraction can often be detained from the intensity I_n of a resultant narrow peak in the long slit angular distribution (Kerr et al. 1965).   The annihilation of positrons that have not formed positronium will contribute a further component or components.

This simple interpretation will not be adequate in the presence of ortho$\rightarrow$para conversion or more complex chemical effects.   In addition, a misinterpretation of the spectrum may easily result from an anomalously slow decay of ' free ' positrons or the ultra-fast pick-off of some or all of the positronium atoms.   However, further checks on the orthopositronium fraction can be obtained from magnetic quenching or measurement of the ratio of three-photon to two-photon events.

In a magnetic field (Halpern 1954 ; Wallace 1960, Goldanskii 1967) the singlet and triplet states are mixed.   At sufficiently large fields the anni-hilation mode is determined by the magnetic quantum number m.   States with $m = \pm 1$ decay with three-photon emission, and those with $m = 0$ by two-photon emission.   Since each of these possibilities is represented by a weight $\frac{1}{2}$, the application of a strong magnetic field produces a one-third reduction of the long-lived orthopositronium fraction, and thus provides a convenient method for its identification.

The three-photon rate can be assessed by a triple coincidence technique (De Blonde et al. 1972) or from a study of the energy spectra of the annihilation photons (Gainotti et al. 1964) although neither technique is very precise.   Then, if our interpretation of I_3 is correct, it is easy to show that the probability of three-photon emission (when small)

$$P_{3\gamma} = \frac{I_3 \tau_3}{t_{\text{ortho}}} + (1 - \tfrac{4}{3}I_3)/372$$

An explanation of the actual intensity or lifetime of any component demands, in general, consideration of the chemical, electronic, and structural properties of the system.   Where there are molecules or un-paired electrons that react strongly with a positronium atom, inhibition and quenching effects may be dominant in determining the form of the lifetime spectrum.   A few examples will be sufficient to illustrate the consequent complexity of positron behaviour in some materials.

The three lifetime components that are generally observed in condensed gases may be tentatively attributed to parapositronium decay, annihilation of free positrons, and pick-off annihilation of orthopositronium. A consistent picture of the positronium fraction in helium is obtained from both the intensity of the longest-lived lifetime component and that of the narrow component in the angular distribution (Stewart and Briscoe 1967). The identification of the positronium components allows a fairly complete description of the positron states (§ 2.2.2).

The interpretation of lifetime spectra and angular distributions in other condensed gases has been less successful. Angular correlation studies (Briscoe *et al.* 1968) directed at the measurement of the parapositronium peak (fig. 7) provide values for the parapositronium fraction approximately three times larger than that inferred from lifetime studies (Paul 1958, Liu and Roberts 1963). The effect appears to depend on low levels of oxygen

Fig. 7

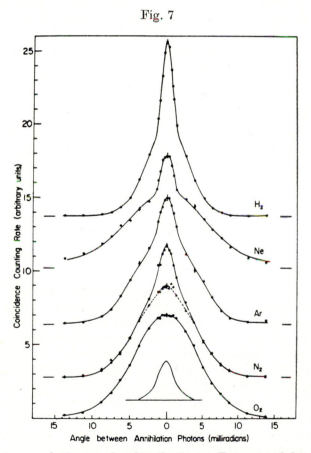

Parapositronium peaks in angular distributions. Two sets of data are shown for liquid nitrogen : Solid line 99·998% purity, broken line 99·5% purity (Briscoe and Stewart 1967).

impurity.   Recent lifetime studies in argon (Spektor *et al.* 1971) and nitrogen (Spektor and Paul 1971) show that the oxygen concentration dependence of the long-lived lifetime component is complex and the variety and nature of the underlying processes are yet to be explained.

A marked sensitivity to low levels of oxygen impurity is also found in the lifetime spectra of light organic liquids.   These spectra can be clearly resolved into two components.   Here the intensity of the longer lifetime orthopositronium pick-off component is usually consistent with the positronium fraction deduced from the intensity of the parapositronium peak in the angular distribution and approaches the fraction suggested by an Ore Gap analysis (Kerr *et al.* 1965).   The lifetime τ_2 of this component is however severely affected by the presence of dissolved oxygen whose quenching is believed to depend on the formation of a positronium oxygen compound (Cooper *et al.* 1967).

In solid and liquid organic halides I_2 is reduced.   The effect depends on the halogen atom, increasing through chlorine→bromine→iodine, but showing little dependence on the molecule as a whole (Kerr and Hogg 1962, Germagnoli *et al.* 1966).   The decrease in I_2 points to the inhibition of positronium formation which Goldanskii (1967) has attributed to positron capture with dissociation of the type

$$XCl + e^+ \rightarrow X + e^+Cl.$$

Tao (1970) has confirmed both inhibition and quenching by dissolved iodine in various organic solvents.   A detailed account of these and many other aspects of positronium chemistry has been given by Goldanskii (1968).

In spite of these many chemical effects the physical (rather than chemical) aspects of the problem can often be inferred from a study of the intensity and lifetime of the orthopositronium components.   In such cases the important factor is the creation or prior existence of a favourable environment for a positronium atom.   Thus phase is important.

2.2.2. *Liquids*

We find in the description of positron lifetimes in liquid helium the basis for the interpretation of positron phenomena in more complex liquids.

The intermediate lifetime in liquid He (1·9 nsec) is too short to be explained by a simple polarization of the surrounding atoms by the positron (Ore 1949).   More sophisticated models explain the lifetime in terms of a positron helium atom bound state with consequent condensation of surrounding atoms around the ion (Roellig 1967).   Similarly, the long lifetime (88 nsec) is too long to be associated with simple pick-off of orthopositronium atoms residing in the bulk material (Ferrell 1957).   This component is now attributed to the annihilation of orthopositronium existing in a bubble, created by repulsive exchange forces between the orthopositronium and helium atoms (Ferrell 1957).   The total energy of the bubble consists of the zero point energy of the positronium atom, a pressure volume term, and a surface energy contribution, i.e.

$$E = E_0(R) + \tfrac{4}{3}\pi R^3 p + 4\pi R^2 \sigma, \tag{19}$$

where p is the pressure, σ the surface tension and R the bubble radius. A minimization of the total energy with respect to R results, at normal pressures, in an equilibrium bubble radius of 10 to 20 Å. A calculation of the annihilation rate resulting from normal orthopositronium decay within the bubble and pick-off annihilation from 'leakage' of the orthopositronium atom into the bubble wall provides values ranging from 78 to 89 nsec which compare well with the experimental value (Roellig 1967). An equally successful theory of the pressure dependence of the width of the parapositronium peak in the angular distribution can be obtained with the bubble model (Hernandez and Choi 1969).

A quantitative analysis of pick-off decay in more complex systems is difficult. Bisi *et al.* (1970) have investigated the decay rate of orthopositronium in a variety of carefully degassed molecular liquids. Their analysis suggests that each atom contributes to the quenching with an effective number of electrons which is independent of the structural form of the molecules. A rather different approach has been adopted by Gray *et al.* (1968) who, demonstrating remarkable perseverance, have measured the lifetime spectra for 193 organic liquids. Their results establish a correlation between the orthopositronium pick-off rate and both the structural form and electronic polarizability of the molecules. Tao (1972) has used these and other results to deduce a correlation between pick-off rate and surface tension. A theoretical basis for this correlation is again developed from the bubble model. The dependence upon the surface tension then provides for other simple relationships between the annihilation rate and properties such as polarizability, cohesive energy density, as well as pressure and temperature changes.

2.2.3 *Solids*

A description of the pick-off decay of orthopositronium atoms is also valuable in the case of molecular solids in which positronium atoms may be confined in already existing holes or cavities. In this respect molecular solids are 'defected' and the present section might well have been included in chapter 4. The most interesting effects arise in polymers for which, in recent years, developments in timing and data analysis have revealed the existence of at least three lifetime components. An illuminating interpretation of these components has been given by Brandt and Spirn (1966) in an analysis of the spectra for polyethylene, polytetrafluoroethylene and glycerol. Since their analysis contains an important example of the use of the statistical models discussed in chapter 4, we shall follow it in some detail.

A study of the temperature and volume dependence of the intensity of the various components (fig. 8) suggests the following picture. The third (longest lifetime) component, whose intensity, I_3, increases with increase of temperature or volume, arises from orthopositronium pick-off in low density regions of the material. I_1 decreases with increasing temperature and is always larger than $I_3/3$ .Consequently I_1 must be the sum of at least three components.   One results from the normal decay of parapositronium.

Fig. 8

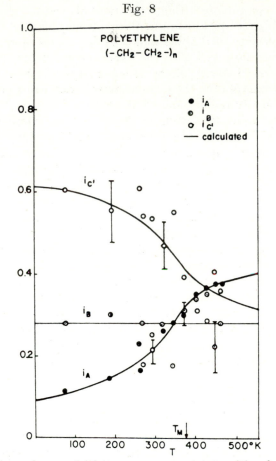

Temperature dependence of lifetime component intensities for polyethylene. i_{C1}, i_B, i_A are respectively I_1, I_2, I_3 in this text (Brandt 1967).

The second emanates from fast pick-off of parapositronium located in regions of high electron density.   A third will arise from the similar fast decay of orthopositronium. I_2, which shows little dependence on temperature or volume, is tentatively assigned to the annihilation of positrons that have not formed positronium.

The temperature dependence of the pick-off annihilation modes is explained with the aid of a ' free volume ' model (Brandt et al. 1960).   The positronium atom may be regarded as moving in a potential distribution which increases rapidly at the boundary of each molecule (Wallace 1960). If the positron is excluded from some volume V around each molecule the average reduced free volume per molecule is

$$v = (V_0 - V)/V_0.$$

V_0 is the volume of a molecular cell. Pick-off annihilation will arise from the penetration of the bound positron wavefunction into the surrounding molecules and its rate will decrease with increase of free volume. Lagu *et al.* (1969) and Thosar *et al.* (1969) use such a model to explain correlations between pick-off annihilation rate and intensity in various materials.

An increase in I_3, obtained at the expense of I_1, implies a transfer of positronium atoms from small free volume to large free volume regions. Brandt and Spirn argue that for volumes above a certain critical volume, the material breaks up into ordered regions of free volume v_c and intermolecular regions of larger free volume. Simple statistical arguments (Turnbull and Cohen 1961) suggest that, if v_f is the average free volume per molecule, the probility of a free volume greater than v_c is given by

$$P_c = \exp(-v_c/v_f). \tag{20}$$

We now assume that n_0 orthopositronium atoms, once formed, are found in intermolecular regions with a probability P_c and sample many such regions during their average lifetime λ_2^{-1}. Alternatively they may, with probability $(1 - P_c)$, find themselves in ordered regions, in which they annihilate at a rate λ_1. The total orthopositronium population then decays with time according to

$$n(t) = n_0 P_c \exp(-\lambda_1 t) + n_0(1 - P_c) \exp(-\lambda_2 t). \tag{21}$$

We may associate the second term with the longest-lived component I_3, τ_3 and the first with the changing contribution to I_1, τ_1. The curves of fig. 8 are based on density temperature curves and the choice $v_c = 0.3$.

The temperature dependence of the lifetime τ_3 in polythene and Teflon is shown in fig. 9. Uncertainties of experiment and statistical analysis prevent a reliable assessment of corresponding changes in the other components although there is some indication of little change in these quantities. The general trends in fig. 9 can be understood in terms of increasing free volume resulting from the decrease in density on heating. At higher temperatures the effects of lattice vibrations and density fluctuations may become important. The basic effect of thermal motion of the molecules is to raise the effective electron density in the free volume, thus decreasing τ_3 relative to its value in a cold lattice of the same density (Brandt 1967). In some cases this effect can outweigh that arising from the increase in volume, such that a maximum develops. A theoretical description of these effects can be obtained from a model in which the molecules, considered as hard-core repulsive potentials of various geometrical forms, move with mean square displacements given by the

R. N. West

Fig. 9

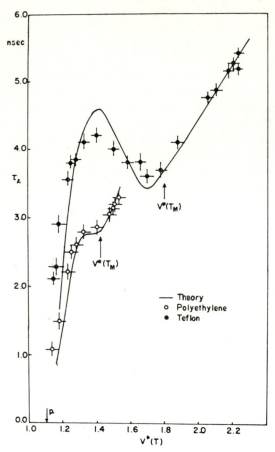

Temperature dependence of experimental and theoretical pick-off lifetimes. The dependence is displayed here through the reduced free volume $v^*(T)$ (Brandt 1967).

Lindemann melting criterion (Pines 1963). The pick-off annihilation rate can then be expressed in terms of functions that depend on the hard-core potential and the size and form of the free volume (Brandt and Spirn 1966, Brandt and Fahs 1970). Numerical tabulations of these functions are available (Brandt and Fahs 1970). The solid curves of fig. 9 pertain to a cylindrical cell geometry. Although the theory has the ability to reproduce the observed variation in lifetime it does not attempt to account for the absolute rates.

An interpretation of the intermediate component I_2, τ_2 in terms of the annihilation of free positrons differs from that of Tao and Green (1964) who have attributed it to the annihilation of orthopositronium in crystalline regions of the material. Various attempts to resolve this question have been made with simultaneous lifetime–momentum measurements. The typical momenta associated with the pick-off annihilation of ortho-positronium atoms should be broader than that for the annihilation of free positrons because of a momentum contribution from the internal motion of the positronium atom. The Tao–Green hypothesis would suggest that the typical momenta associated with I_2 should be at least as large as those associated with I_3. Combined Doppler shift lifetime measurements suggest that the momentum distributions are similar (I. K. MacKenzie, private communication). However, simultaneous lifetime angular correlation studies (Hsu and Wu 1967, McGervey and Walters 1970) indicate a relative enhancement of I_3/I_2 for high momentum events. Although the latter measurements appear to rule out the Tao–Green hypothesis they are not sufficient to confirm the picture in which I_2 results from the annihilation of free positrons. It is conceivable that some or all of the intermediate components originate from positrons bound to the molecules of the medium (McGervey and Walters 1970). This interpretation would explain the observed temperature or volume insensitivity of I_2, τ_2. Despite the confusion Cova and Zappa (1968) have established a correlation between τ_2 and the size and electronic structure of the molecular atoms in various materials.

A quantitative theory of relative fraction and lifetime of any of the resolvable positron states in molecular solids would seem at the present to be remote. Studies of variations in intensity of the orthopositronium pick-off component would certainly seem to offer the best chance of experimental studies of, if nothing else, the microstructure of the system.

2.2.4. *Phase transitions*

Investigations of the effects of melting on positron parameters provide a good test of the validity of the models discussed in previous sections. A recognition of the importance of structure as a factor affecting the formation and annihilation of positronium is apparent in many early studies of solid–solid and solid–liquid phase transitions (Wallace 1960, Goldanskii 1967). Long lifetime components observed in the spectra for plastic sulphur and fused quartz were not discernible in the spectra for the crystalline forms (Bell and Graham 1953). The melting of napthalene (Landes *et al.* 1956) is accompanied by an increase in the intensity of the long-lived component. In some cases (Page and Heinberg 1956, De Zafra and Joyner 1958) angular correlation studies confirm the association of the long-lived component with orthopositronium decay. These results are, at least superficially, in accord with the role of order in positronium formation and stability as described by the qualitative arguments of the previous sections and the modified Ore Gap analysis discussed in § 2.3 to follow. Such is not always the case.

A good understanding of positron behaviour in water is a basic requirement for the adequate interpretation of aqueous solution positron chemistry (McGervey 1967, Green and Lee 1964). Early studies of the water–ice transition, which has received a great deal of attention, were hard to explain. A two-component analysis of the lifetime spectra (Fabri *et al.* 1963) showed a relatively smooth increase in τ_2 and decrease in I_2 on melting. Studies of the parapositronium peak in the angular distributions (De Zafra and Joyner 1958, Colombino *et al.* 1965), magnetic quenching (Fabri *et al.* 1967) and the three-gamma annihilation rate (Wagner and Hereford 1955) gave conflicting pictures of the relative positronium yields (Goldanskii 1967).

More recent studies by Eldrup *et al.* (1971), although emphasizing the difficulties that can arise in the identification of orthopositronium lifetime components, provide an interpretation roughly in line with the ideas of previous sections. Two or three-component analyses of the spectra for water result in an intensity ($\sim 27\%$) and lifetime ($\sim 1 \cdot 9$ nsec) for the orthopositronium pick-off component substantially in agreement with the values deduced by earlier workers. In the solid the picture is more complex. Two and three-component analyses of spectra obtained at sample temperatures $-200°C \rightarrow 0°C$ suggest a complex pattern of variations in the intensities and lifetimes of the longer-lived components. As the temperature is raised, the intensity, I_3, of the slowest component (as deduced from a three-term analysis) increases at the expense of the intermediate component, I_2. The pattern is similar to that described in § 2.2.3 in connection with the lifetime spectra of polymers. A similar theoretical analysis associates the increase in I_3 with orthopositronium trapping in temperature-induced defects. The trapping centres are not identified but the addition of a small concentration of defect-creating hydrogen fluoride impurities results in the emergence of a long lived component of lifetime $1 \cdot 23$ nsec and intensity $\sim 52\%$. As a result of this and other considerations, including the results of angular correlation measurements of the orthopositronium fraction (Mogensen *et al.* 1971) and (privately communicated) magnetic quenching studies, Eldrup and his colleagues conclude that the total amount of orthopositronium formed in ice is $\sim 52\%$ independent of temperature. Spectrum analyses based on a two state trapping model (§ 4.1) provide a plausible description of the spectra for the HF doped samples. A precise description of the origins of the longer lived components in the spectra for the nominally pure samples is more difficult but there seems little doubt that orthopositronium trapping by defects still plays an important role. The somewhat longer orthopositronium lifetime in water again can be explained by the bubble model discussed in § 2.2.2.

Similar changes in lifetime spectra, on melting of D_2O (Eldrup *et al.* 1971), CH_3OH and CH_3OD (Chuang and Tao 1970) have been reported. However, the melting of some hydrocarbons produces a rather different pattern (De Blonde *et al.* 1972). In liquid CH_4, C_6H_6, and C_6H_{12} the expected ratio, $I_n = I_2/3$, between the intensity of the narrow component in the regular distribution, I_n, and that of the long lifetime component I_2, gives an acceptable picture of positronium fraction. In the solids there is a marked disagreement between $I_2/3$ and I_n which is essentially zero and suggests that no positronium is formed. Measurements of the triple coincidence rates again suggest that little or no positronium is formed. A convincing explanation of these results has yet to be given.

Changes in positron parameters at solid–solid transitions have also been observed (Goldanskii 1967). Cooper *et al.* (1969) report a dramatic change in the lifetime of a long-lived spectrum component at a solid–solid transition in cyclohexane and suggest that positron studies provide a rapid method of surveying for solid–solid phase transitions. Phase transitions in polymers have also been investigated (Goldanskii 1967, Stevens and Mao 1970). In some cases the measurement of a transition temperature different to that obtained by other techniques poses new questions about the nature of the transition (Chuang *et al.* 1972).

2.3. *Ionic crystals*

The criterion for positronium formation attributed to Wallace (1960) in § 2.1 was based on the experimental results available at the time which gave no evidence of positronium in ionic crystals. Ferrell (1956) had also shown, from a modified Ore gap analysis, that positronium formation was extremely unlikely.

In an ordered solid having energy bands, the role of the excitation energy in expression (17) is taken over by the energy required, V_1, to promote an electron to the unoccupied state of lowest energy. In such materials the lowest energy positron and positronium states are also important. If the positron affinity of the crystal, i.e. the energy required to remove a thermalized positron from the system, is denoted by Q_p, and that for the positronium atom by Q_{pos}, the inequality becomes,

$$V_1 > E > (V_i + Q_p) - (\text{B.E.} + Q_{pos}).\tag{22}$$

B.E. is the binding energy of the positronium atom. Alternatively, if $Q_e = (V_i - V_1)$ is the electron affinity of the crystal, the criterion for the existence of the Ore gap may be written

$$Q_e + Q_p < Q_{pos} + \text{B.E.}\tag{23}$$

Thus positronium formation will be inhibited if the combined positron and electron affinities are large. Naïve considerations suggest that the positron will exist mainly in the outer electron shells of the negative ions. With this assumption Ferrell deduced that the positron affinity is small and that the sum of the positron and electron (Seitz 1964) affinities is only of the order of 1–2 eV.

An assessment of the right-hand side of expression (21) is more significant. If positronium exists in a perfect ionic crystal it must be interstitial. The interstitial space is however limited and if the positronium atom is confined in a space of less than its normal dimensions the binding energy will be reduced, the limiting case resulting in no bound state remaining.   In addition, the zero point motion of the confined atom of light mass will contribute additional kinetic energy which Ferrell estimates to be more than 10 eV.   Thus it would appear that the right-hand side of expression (21) is likely to be large and negative and more than sufficient to destroy the Ore Gap.   Nevertheless we should note that the strength of this argument depends on the non-availability of space for the atom and should space be provided, as for example by the presence of point or extended defects, the picture may be radically altered.

Early measurements of positron lifetimes in ionic crystals suggested a single lifetime of the order of 200 picoseconds (Bell and Graham 1953). Angular correlation data (Lang and de Benedetti 1957) appeared consistent with a picture of the positron in a bound s-state about a negative ion and stimulated several calculations of the binding energy and annihilation rate for such states (Neamtan et al. 1962, Neamtan and Verral 1964). Angular correlation studies in alkali halides (Stewart and Pope 1960) confirmed the picture of positrons annihilating with the outer electrons of negative ions but were also consistent with a positron wavefunction that was almost a plane wave and not localized on a single ion.   Difficulties involved in the interpretation of the angular distribution for lithium hydride (Stewart and March 1961) in terms of delocalized positron states were eventually shown (Brandt et al. 1966) to arise from an inadequate description of the states of participating electrons.

An indication that delocalized positron states do not exhaust the possibilities was provided by the discovery of long lifetime components in the spectra of alkali halides (Bisi et al. 1963).   Further investigations revealed that the magnetic quenching (§ 2.1) was too small (Bisi et al. 1964) and the ratio of three-photon to two-photon events too large (Gainotti et al. 1964, Bisi et al. 1966) for the long-lived component to originate from the decay of a simple positronium atom.   More refined experiments by Bussolati et al. (1967) revealed the presence of at least three components in some materials.   These workers suggested that the main components could arise from the decay of the ground and first excited states of a positron-negative ion bound state, the existence of which had been postulated by Goldanskii and colleagues (Goldanskii and Prokop'ev 1965, 1966, Prokop'ev 1966).   Their additional postulate of positrons localized in defect centres can reasonably be said to have stimulated the avalanche of positron defect studies reported in chapter 4 of this article.   A short but valuable list of annihilation or A-centres has been given by Brandt (1967) and is reproduced as table 1.   The existence of some of these states has now been demonstrated in many experiments (§ 4.2), and the possibility of positronium and positron-negative ion compounds (Wheeler compounds) has prompted renewed interest in the theoretical treatment of positron bound states with both electrons (Ferrante and Geracitano 1970, Stancanelli and Ferrante (1970) and ions (Schrader 1970).

Table 1.   The simplest A centres in ionic crystals (Brandt 1967)

Designation†	Name	Lifetime	Angular correlation
$[+ \mid e^+ \mid \quad]$	Vacancy A centre	$\sim$ Eqn. (16)	As crystal
$[+ \mid e^+e^- \mid \quad]$	Vacancy A_-' centre	$\sim p_s$	Narrow component
$[- \mid e^-e^+ \mid \quad]$	Vacancy A_+' centre	$\sim p_s$	Narrow component
$[\quad \mid e^+ \mid -]$	Interstitial A centre	$\sim$ Wheeler compound	

† Abbreviation for [missing ion|trapped particles|added ion].

2.4. *Metals*

2.4.1. *The free-electron approximation*

The theoretical arguments against the formation of positronium in perfect ionic solids are equally applicable to metallic systems. The screening effect of a dense electron gas provides a further reason to doubt the possibility of positron binding to a single electron. The experimental lifetime spectra for most metals appear to consist of a single component thus implying a common state for the annihilating positrons (fig. 6).

The positrons move in an array of repulsive ion potentials and the best calculations (Hodges 1970) suggest that thermalized positrons will annihilate from a quasi-free state close to $(\sim k_B T)$ the bottom of a positron conduction band. Since Coulomb repulsion will tend to exclude the positron from regions well inside the ion cores it will annihilate in the main with conduction electrons. It is thus instructive to compute the annihilation parameters according to the Sommerfeld model.

Doubly-occupied spin independent free electron states,

$$\psi_\mathbf{k}(\mathbf{r}) = V^{-1/2} \exp(i\mathbf{k}_- \cdot \mathbf{r}),$$

define a spherical Fermi surface, $k_- \leqslant k_F$, the Fermi momentum. In the same spirit we define a zero temperature state for the thermalized positron $(\mathbf{k}_+ = 0)$ by $\psi_+(\mathbf{r}) = V^{-1/2}$. In terms of these functions, eqns. (15), (16), for the photon pair momentum distribution, have the form

$$\Gamma(\mathbf{k})d^3\mathbf{k} \propto \sum_{k_- \leqslant k_F} \delta(\mathbf{k} - \mathbf{k}_-) \, d^3\mathbf{k}. \qquad (24)$$

The conventional long slit angular correlation apparatus measures a quantity proportional to the distribution of one component of the pair momentum $\mathbf{p} = \hbar\mathbf{k}$, that is,

$$N(p_z) \, dp_z \propto \int_{-\infty}^{+\infty} \int_{-\infty}^{+\infty} \Gamma(\mathbf{p}) \, dp_x \, dp_y \, dp_z. \qquad (25)$$

For an isotropic distribution, $\Gamma(p)$ (Stewart 1957),

$$N(p_z) \, dp_z \propto 2\pi \int_{p_z}^{\infty} p\Gamma(p) \, dp \, dp_z \qquad (26)$$

which, in the present case, provides a parabolic distribution,

$$N(p_z)\, dp_z \propto 2\pi(p_F^2 - p_z^2)\, dp_z \quad (p_z < p_F) \Big\}$$

$$N(p_z)\, dp_z = 0 \qquad\qquad (p_z > p_F) \Big\} \tag{27}$$

A parabolic component is observed in the angular distributions of many metals, and the ' cut-offs ' at $p_z = p_F$ are in good agreement with the predictions of the Sommerfeld theory applied to the conduction electron density (Stewart 1957, Lang and de Benedetti 1957). There is in addition a broader component of roughly gaussian form which arises both from more subtle features of the conduction electron structure and annihilation with more tightly bound electrons. The latter effect, usually referred to as core annihilation, is now believed to provide the major contribution to any significant broad component (Berko and Plaskett 1958, Gould *et al.* 1972). In some metals, which we shall henceforth refer to as simple, the broad component is of small intensity and it would seem that conduction electrons account for almost all of the annihilation events.

Table 2. Positron lifetimes in metals at or near room temperature. (1) Weisberg and Berko (1967) ; (2) Mackenzie *et al.* (1967) ; (3) Hautojärvi *et al.* (1970) ; (4) McKee, Jost and Mackenzie (1972) ; (5) Cotterill, Petersen, Trumpy and Träff (1972) ; (6) Mackenzie, Craig and McKee (1971) ; (7) Hautojärvi and Jauho (1971) ; (8) Cotterill, Mackenzie, Smedskjaer, Trumpy and Träff (1972) ; (9) Crisp *et al.* (1973) ; (10) West *et al.* (1973). The generally lower values of τ_1 obtained from more recent work result from the recognition of defect-trapping phenomena (§ 4.3).

metal	τ_1(psec) (main component)	Ref.	metal	τ_1(psec) (main component)	Ref.
Li	291 ± 6	1	Zn	179 ± 5	2
Li	301 ± 2	6	Ga	194 ± 4	1
Be	213 ± 5	1	Rb	406 ± 10	1
Na	338 ± 7	1	Rb	425 ± 2	6
Na	344 ± 2	6	Mo	146 ± 3	8
Mg	232 ± 5	1	Cd	185 ± 4	1
Al	201 ± 5	1	Cd	196 ± 3	2
Al	175 ± 4	2	In	203 ± 5	2
Al	172	3	In	188 ± 12	9
Al	174	4	Sn	202 ± 5	1
Al	166 ± 2	5	Cs	418 ± 10	1
K	397 ± 10	1	Hg(liq.)	220 ± 7	1
K	400 ± 1	6	Hg(sol.)	185 ± 10	10
Fe	160 ± 5	1	Pb	201 ± 5	1
Co	162 ± 4	1	Bi	243 ± 8	1
Ni	172 ± 5	1			
Cu	181 ± 5	1	Si	222 ± 5	1
Cu	147 ± 3	2	Ge	226 ± 5	1
Cu	132 ± 7	7			

This apparently successful interpretation of the angular distributions for simple metals by a model which completely neglects the strong Coulomb attraction between the positron and electrons is at first sight surprising. The inability of the Sommerfeld model to produce a correct picture of the real physical situation is however clearly demonstrated by the measured total annihilation rates or reciprocal lifetimes (table 2). These rates are considerably greater and show less dependence on the average electron density than calculated Sommerfeld rates (fig. 10). An explanation of how these effects arise from the response of the interacting electron gas to the positron force has required, and provided a demanding test of the modern techniques and concepts of many electron physics.

Fig. 10

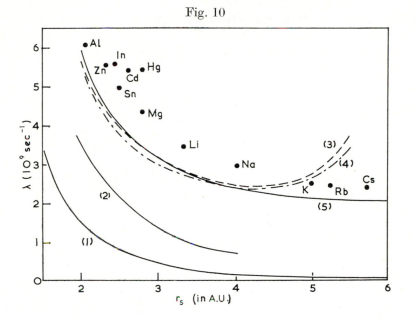

Experimental and theoretical electron gas annihilation rates for metals. Theoretical curves : (1) Sommerfeld model. (2) High density approximation (Kahana 1963). (3) Ladder approximation (Kahana 1963). (4) Sjölander and Stott (1972). (5) Bhattacharyya and Singwi (1972 a). The experimental data are taken from table 2.

2.4.2. *Electron gas theories*

The polarization effects which determine positron lifetimes in metals are more complex than that of the simple screening of a static point impurity. Because of the small positron mass a really adequate treatment of the positron problem must take account of the correlated motion of the interacting particles. Such correlations will also strongly influence the thermalization and equilibrium behaviour of positrons in metals. Sophisticated calculations by Carbotte and Arora (1967) and Woll and

Carbotte (1967), although not in exact agreement with the earlier results of Lee-Whiting (1955), still produce thermalization times that are, in the most part, significantly shorter than the corresponding lifetimes. For the present we defer consideration of the details of these analyses and merely invoke their general result to suggest that calculations of the more readily observable lifetime may be plausibly developed from an initial state of zero positron momentum.

Some of the most successful and conceptually useful calculations make use of the close relationship between the annihilation rate, as given in eqn. (13), and the zero temperature electron–positron Green's function or propagator

$$G_{ep}(x, y \; ; \; x', y') = (-i)^2 \langle T\{\psi_-(x)\psi_+(y)\psi_+^\dagger(y')\psi_-^\dagger(x')\}\rangle. \tag{28}$$

Here, the $\psi(x)$ are Heisenberg field operators of argument $x = \mathbf{r}, t$, etc., T is the Wick time-ordering operator (Wick 1950), and the expectation value is taken with respect to the fully interacting ground state of the positron–many-electron system.

The total annihilation rate may then be expressed in terms of G_{ep}, and the electron density, n_0, and annihilation rate, Γ_0, of parapositronium (Kahana 1963, Carbotte 1967) by the equation

$$\Gamma = i^2(\Gamma_0/4n_0) \int d^3\mathbf{r} \; \underset{t' \to t_+}{\text{Lim}} \; G_{ep}(\mathbf{r}t, \mathbf{r}t \; ; \; \mathbf{r}t', \mathbf{r}t'), \tag{29}$$

where the limiting process ensures the correct ordering of the positron and electron field operators.

The equations of motion for the interacting system couple G_{ep} to similar functions involving both fewer and greater numbers of particles. More conveniently one may argue a Bethe–Saltpeter type of integral equation for G_{ep} (Kahana 1963, Ziman 1969)

$$G_{ep}(x, y \; ; \; x', y') = G_e{}^0(x, x)G_p{}^0(y, y') + (-i) \int d^4\xi \, d^4\eta \, d^4\xi' \, d^4\eta'$$

$$\times \, G_e{}^0(x, \xi)G_p{}^0(y, \eta)I(\xi, \eta \; ; \; \xi'\eta')G_{ep}(\xi', \eta' \; ; \; x', y'). \tag{30}$$

Here

$$G_e{}^0(x, x') = (-i)\langle T\{\psi_-(x)\psi_-^\dagger(x')\}\rangle_0 \left.\begin{array}{c} \\ \\ \end{array}\right\}$$

$$G_p{}^0(y, y') = (-i)\langle T\{\psi_+(y)\psi_+^\dagger(y')\}\rangle_0 \tag{31}$$

are the free particle electron and positron Green's functions for the non-interacting system.

The first term on the right-hand side of eqn. (30) gives the simple Sommerfeld result $\Gamma^{(0)}$. To proceed further one requires an expression for the interaction $I(\xi, \eta \; ; \; \xi'\eta')$ which, in principle, itself is a function of G_{ep} and other propagators involving arbitrary numbers of particles. Thus approximations are necessary.

Any realistic approximation must acknowledge the modification of the direct Coulomb force between the positron and a particular electron by the screening of the surrounding electron gas medium. It is known (Hubbard 1957) that some measure of the screening is obtained by a summation, in the random phase approximation (R.P.A.), of the simple polarization or bubble diagrams (fig. 11). This sum which represents the

Fig. 11

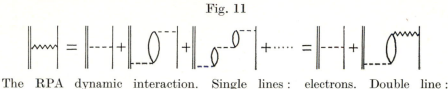

The RPA dynamic interaction. Single lines: electrons. Double line: positron.

excitation of an arbitrary number of electron hole pairs in the background gas can be evaluated by the usual rules (Pines 1962, Mattuck 1967) to provide another integral equation, this time for an effective interaction (Kahana 1963):

$$u(x-y) = v_{ep}(x-y) + (-i) \int d^4z \, d^4z' \, v_{ee}(x-y)$$

$$\times G_e^0(z, z') G_e^0(z, z') u(z'-y). \quad (32)$$

$v_{ep} = -v_{ee}$ is the bare Coulomb potential and the interaction

$$I(x, y \; ; \; x', y') = u(x-y) \delta^4(x-x') \delta^4(y-y'). \quad (33)$$

With such a two-point interaction eqn. (30) represents the sum of the so-called ladder diagrams (Prange and Klein 1958). Some low-order ladder diagrams are shown in fig. 12.

Metallic conduction electron densities, ρ_0, (fig. 10) lie in the range $1 \cdot 8 \leqslant r_s \leqslant 5 \cdot 5$ A.U. where r_s is the usual parameter $r_s = (\frac{4}{3}\pi\rho_0)^{-1/3}$. At the highest densities one can expect the first iteration of eqn. (30) to provide a reasonable approximation (Pines 1962), i.e.

Fig. 12

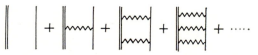

The ladder sum. Single line: electron. Double line: positron.

$$G_{ep}(x, y \; ; \; x', y') = G_e^0(x, x') G_p^0(y, y') - i \int d^4x'' \, d^4y''$$

$$\times G_p^0(y, y'') G_e''(x, x'') u(x''-y'') G_e^0(x'', x') G_p^0(y'', y'). \quad (34)$$

The evaluation of eqns. (32), (34) is carried out in a momentum-energy representation. For a spatially homogeneous system of volume

V, $G^0(x, x') = G^0(x - x')$, and we may define a Fourier transform,

$$F(x - x') = \frac{1}{2\pi V} \sum_{\mathbf{k}} \int d\omega \exp\left[i\mathbf{k} \cdot (\mathbf{r} - \mathbf{r}') - i\omega(t - t')\right] F(\mathbf{k}, \omega). \qquad (35)$$

Of particular importance are the free-particle functions (Pines 1962)

$$G_e^0(\mathbf{k}, \omega) = \frac{\theta(k - k_F)}{\omega - k^2 + i\delta} + \frac{\theta(k_F - k)}{\omega - k^2 - i\delta}, \qquad (36\,a)$$

$$G_p^0(\mathbf{k}, \omega) = \frac{\theta(k)}{\omega - k^2 + i\delta} + \frac{\theta(-k)}{\omega - k^2 - i\delta}. \qquad (36\,b)$$

$$(\hbar/2m = 1, \quad \delta = O^+)$$

The second equation requires that the initial state contain one positron of zero momentum. Fourier transformation of the second term of eqn. (34) then yields, after various algebraic steps, a correction to the Sommerfeld rate,

$$\Gamma^{(1)} = (i\Gamma_0/\pi n_0 V) \sum_{\mathbf{q}} \int d\omega\, u(\mathbf{q}, \omega) Q(\mathbf{q}, \omega) \left\{\frac{1}{\omega - q^2 + i\delta}\right\}. \qquad (37)$$

The interaction $u(\mathbf{q}, \omega)$, can be obtained from a similar treatment of eqn. (32), whence

$$u(\mathbf{q}, \omega) = -v_q/\epsilon(\mathbf{q}, \omega), \quad \epsilon(\mathbf{q}, \omega) = 1 + 2v_q Q(\mathbf{q}, \omega), \qquad (38)$$

where $v_q = 4\pi e^2/q^2$ is the Fourier transform of the Coulomb potential. The properties of the polarization part $Q(\mathbf{q}, \omega)$ are well established (Dubois 1959), and $u(\mathbf{q}, \omega)$ contains contributions from both short-range screening and plasmon excitation. The presence of $Q(\mathbf{q}, \omega)$ in both eqns. (37) and (38) emphasizes the dual role of the polarization in both screening the direct interaction and enhancing the annihilation rate.

If $Q(\mathbf{q}, \omega)$ is written in full.

$$\Gamma^{(1)} = \frac{i\Gamma_0}{\pi n_0 V^2} \sum_{\mathbf{p}} \sum_{\mathbf{q}} \int d\omega\, \frac{u(\mathbf{q}, \omega)}{\omega + q^2 - i\delta}$$

$$\times \left[\frac{\theta(|\mathbf{p}+\mathbf{q}| - p_F)\theta(p_F - p)}{\omega + p^2 - (\mathbf{p}+\mathbf{q})^2 + i\delta} - \frac{\theta(p_F - |\mathbf{p}+\mathbf{q}|)\theta(p - p_F)}{\omega + p^2 - (\mathbf{p}+\mathbf{q})^2 - i\delta}\right]. \qquad (39)$$

By omitting the $\mathbf{p}$ summation one obtains the corresponding contribution $\Gamma^{(1)}(\mathbf{p})$ to the annihilation rate into momentum $\mathbf{p}$.

A numerical solution of eqn. (37) (Kahana 1963) gives annihilation rates which although significantly greater than the Sommerfeld values still fall short of the experimental rates, particularly at the lower densities (fig. 10). One must assume that at these densities the higher-order ladder diagrams, which represent the 'repeated scattering' of the annihilating pair, provide an important contribution to the annihilation rate.

The evaluation of the complete set of ladder diagrams is made difficult by the frequency dependence of $u(\mathbf{q}, \omega)$. The problem is eased by recourse to a suitable static equivalent. The most popular choice has been the zero frequency limit, $u(\mathbf{q}, 0)$, of the interaction already described, although other forms have been suggested (Zuchelli and Hickman 1964, Crowell *et al.* 1966). A further simplification results from the recognition of the very low density of positrons in any normal experimental situation. Every positron-hole line in a diagram produces a factor (V^{-1} in the present formulation), proportional to the positron density, in the corresponding term in the perturbation expansion of G_{ep}. Thus the most important diagrams are those containing only one such positron-hole line (Majumdar 1965 a)†.

The modifications to eqn. (30) suggested by these arguments then result, after a great deal of tedious manipulation (Carbotte 1966), in an expression for the partial rate into momentum $\mathbf{p}$,

$$\Gamma(\mathbf{p}) = (\Gamma_0/4n_0 V)\theta(p_{\mathrm{F}} - p) \left| 1 + \sum_{\mathbf{m}} \chi(\mathbf{p}, \mathbf{m}) \right|^2. \tag{40}$$

The correction to the Sommerfeld rate is contained in the Bethe–Goldstone amplitude, $\chi(\mathbf{p}, \mathbf{m})$ (Bethe and Goldstone 1957), which satisfies (the inevitable integral equation)

$$\chi(\mathbf{p}, \mathbf{m}) = \frac{\theta(m - p_{\mathrm{F}})}{m^2 + (\mathbf{p} - \mathbf{m})^2 - p^2}$$

$$\times \left[\frac{1}{V} u(\mathbf{m} - \mathbf{p}) + \frac{1}{V} \sum_{\mathbf{q}} u(\mathbf{q})\chi(\mathbf{p}, \mathbf{m} - \mathbf{q}) \right]. \tag{41}$$

The solution of eqn. (41) proceeds by approximation and numerical integration (Kahana 1963, Carbotte 1967). The result may be expressed in terms of an enhancement factor, $\epsilon(p) = \Gamma(p)/\Gamma^{(0)}$, which is conveniently represented, in terms of reduced momenta, $\gamma = p/p_{\mathrm{F}}$, by the equation

$$\epsilon(\gamma) = a + b\gamma^2 + c\gamma^4. \tag{42}$$

The enhancement factors obtained by Kahana (1963) are typical and are represented by the coefficients listed in table 3. The total annihilation rates shown in fig. 10 are, at least for $r_s \leqslant 5$ A.U., in good agreement with measured rates for those metals in which the core annihilation rate is believed to be small. The predicted momentum dependence of the anni- hilation rate, although significant, results in a relatively small modification to the doubly integrated (eqn. (25)) long slit angular distribution. The ' swelling ' of the parabolic component (fig. 13) is easily resolved in con- temporary experiments (Donaghy and Stewart 1967 b, Melngailis and

---

† More refined theories may involve diagrams containing more than one positron-hole line. See Bergersen (1969).

de Benedetti 1966) and its prediction must be regarded as a considerable success for the theory.

Table 3.   Parameters for the momentum dependent enhancement factor
$\epsilon(\gamma) = a + b\gamma^2 + c\gamma^4$ (eqn. (42)) (Kahana 1963)

r_s	a	b	c	Annihilation rate $(10^{-9} \text{ sec}^{-1})$
2	3·480	0·600	0·387	5·930
3	6·172	1·292	0·967	3·239
4	11·225	2·940	2·617	2·609

Various modifications and corrections have been suggested. Alternative static interactions proposed and analysed by Zuchelli and Hickman (1964), Crowell *et al.* (1966) and Arponen and Jauho (1968) produce rates similar to those obtained by Kahana. A comprehensive but complex analysis of the electron–positron interaction has been given by Rajagopal and Majumdar (1970). Corrections to the ladder sum arising from particle–hole interactions have also been studied (Kanazawa *et al.* 1965, Crowell *et al.* 1966, Arponen and Jauho 1968, Carbotte and Kahana 1965) but again produce only minor corrections to the Kahana results.

Fig. 13

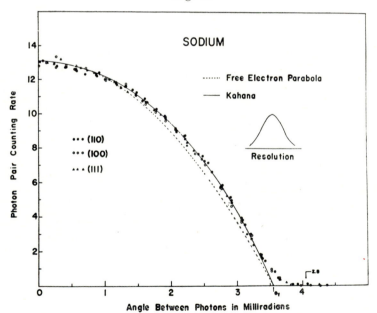

Angular distribution 'swelling' through many-body enhancement (Donaghy and Stewart 1967 b).

The comparative insensitivity of the final results to the detailed choice of interaction lends some support to the use of a simple static potential in the ladder approximation. Further support is provided by the investigation of self-energy corrections made by Carbotte and Kahana (1965).

A first-order calculation of the effect of electron–electron interaction on the equilibrium electron momentum distribution (Daniel and Vosko 1960) results in a distribution having a significant tail for momenta $p > p_f$ (fig. 14). Similarly, positron self-energy effects provide a positron momentum distribution which is non-vanishing for finite momenta. The naive conclusion would be that a similar tail should be present in the pair momentum distribution. There is, however, an additional effect that must be considered.

Fig. 14

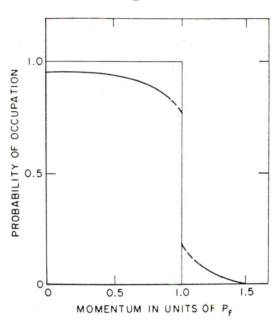

The electron momentum distribution for an interacting electron gas as calculated by Daniel and Vosko (1960) (Kahana 1967).

Contributions to the momentum dependent annihilation rate from the first-order ladder graph (eqn. (39)) can be computed with the full dynamic interaction, $u(\mathbf{q}, \omega)$, or the static version, $u(\mathbf{q}, 0)$, supplemented by a plasmon correction (Kahana 1963). The use of the static potential gives no contribution for $p > p_f$. The dynamic potential results in a more rapid variation in enhancement factor for $p < p_f$, and a *negative* contribution for $p > p_f$. The difference between the static ($+$ plasmon) and dynamic results is almost identical to the modification to the momentum distribution resulting from first order self energy corrections and it would seem (Carbotte and Kahana 1965) that the use of the free-particle propagators is, to a large extent, compensated by the use of the static approximation to the potential.

The exact physical significance of the ' negative contribution ', which arises from the second term in eqn. (39), is not clear to this writer. Nevertheless, we suggest that the following argument goes some way towards explaining the type of cancellation noted above.

The modifications to the electron and positron momentum distributions are caused by the screening cloud of electrons which accompany each particle.   If such modifications are to be included in the theory it is equally necessary to take account of the relaxation of these screening clouds when the electron and positron come together.   This can only be done with a dynamic potential.   Conversely, the use of a static potential and free particle propagators evidently goes some way towards representing the comparatively weaker polarization around an electron–positron pair nearly in coincidence (Hatano et al. 1965).   Whether or not such effects are sufficient to remove all traces of tail in the pair momentum distribution is not clear but Carbotte and Kahana argue the existence of a similar cancellation in all orders of perturbation theory.

A more serious criticism of the ladder approximation was made by Bergersen (1964) who showed that it results in an unphysical accumulation of charge around the positron.   Carbotte (1967), however, was able to demonstrate that this particular defect can be easily repaired by the inclusion of a further set of diagrams which nevertheless have little effect on the annihilation rate.   The criticism and the answer both provide a clear indication of the danger inherent in partial summations of perturbation expansions (Pines 1962).   A rapid and implausible increase in the theoretical annihilation rate for $r_s > 5$ A.U. (fig. 10) again points to the limitations of the ladder theory at lower electron densities.

A general objection to the ladder approximation arises from its inability to do justice to the necessary close relationship between the displaced charge around the positron and the screening of the positron–electron interaction (West 1971).   The high density approximation (eqns. (37), (38)) is better in this respect.   Furthermore, the simple analogy between positron–electron interaction and electron–electron interaction, which leads to eqn. (32), is probably insufficient to expose the proper relationship between these potentials (Rajagopal and Majumdar 1970).

Both of these objections are largely answered by positron lifetime calculations (Sjölander and Stott 1970, 1972, Bhattacharyya and Singwi 1972 a) which make use of the dielectric screening theory of Singwi et al. (1968).   In this theory, the dielectric function for electron–electron interaction is made a functional of the structure factor or Fourier transform of the static pair distribution function.   A self-consistent calculation of the structure factor and dielectric function can then be made with the aid of the fluctuation–dissipation theorem (Pines 1963).   The extension of the theory to a two-component (positron–electron) system has been described by Sjölander and Stott (1972) whose computed positron annihilation rates, shown in fig. 10, are in close agreement with those of Kahana (1963).

Bhattacharyya and Singwi (1972 a) have employed a modification (Vashishta and Singwi 1972) to the theory of Singwi *et al.* (1968) which takes account of the change in a pair correlation function in a weak external field.   A brief account of their work now follows.

The response of a uniform system of electrons of density n_1, and positrons of density n_2 is ' tested ' by the application of weak external potentials $V_1^{ext}(\mathbf{q}, \omega)$ and $V_2^{ext}(\mathbf{q}, \omega)$ (the subscripts 1 and 2 refer, throughout, to electrons and positrons respectively).   The induced charge of the ith component,

$$\langle \rho_i(\mathbf{q}, \omega) \rangle = \chi_i^0(\mathbf{q}, \omega) V_i^{eff}(\mathbf{q}, \omega) \tag{43}$$

where χ_i^0 is the non-interacting response of the ith component, and

$$V_i^{eff}(\mathbf{q}, \omega) = V_i^{ext}(\mathbf{q}, \omega) + \sum_{j=1}^{2} \psi_{ij}(\mathbf{q}) \langle \rho_j(\mathbf{q}, \omega) \rangle. \tag{44}$$

In the theory of Singwi *et al.*, the interaction $\psi_{ij}(\mathbf{q})$ is a functional of $\gamma_{ij}(\mathbf{q})$ the Fourier transform of the pair distribution function $g_{ij}(\mathbf{r}) - 1$.   A relationship suggested by Vashishta and Singwi (1972) for the theory of electron correlation takes the form

$$\psi(\mathbf{q}) = \frac{4\pi e^2}{q^2} + \frac{4\pi e^2}{q^2} \left(1 + an \frac{\partial}{\partial n} \right) \left[\frac{1}{n} \int \frac{\mathbf{q} \cdot \mathbf{q}'}{q'^2} \gamma(\mathbf{q} - \mathbf{q}') \frac{d^3q'}{(2\pi)^3} \right]. \tag{45}$$

A two-component generalization of eqn. (45), together with eqns. (43) and (44), provides expressions for the dielectric functions $\epsilon_{ij}(\mathbf{q}, \omega)$.   Simplified expressions for the case of vanishing positron density, $n_2 \to 0$, then may be used with a generalized fluctuation dissipation theorem

$$n_i[\delta_{ij} + \gamma_{ij}(\mathbf{q})] = -\frac{\hbar q^2}{4\pi^2 e^2} \int_0^\infty d\omega \ \text{Im} \ [1/\epsilon_{ij}(\mathbf{q}, \omega)], \tag{46}$$

to define an integral equation for $\gamma_{21}(q)$:

$$\gamma_{21}(\mathbf{q}) = f(\mathbf{q}) + f(\mathbf{q}) \left(1 + a_{21} n_1 \frac{\partial}{\partial n_1} \right) \left[\frac{1}{n_1} \int \frac{\mathbf{q} \cdot \mathbf{q}'}{q'^2} \gamma_{21}(\mathbf{q} - \mathbf{q}') \frac{d^3q'}{(2\pi)^3} \right]. \tag{47}$$

Here

$$f(\mathbf{q}) = \hbar/\pi m_2 \int_0^\infty d\omega \ \text{Im} \left\{ \left[1/\epsilon_{11}(\mathbf{q}, \omega) - 1 \right] \chi_2^0(\mathbf{q}, \omega) \right\} \tag{48}$$

and

$$\chi_2^0(\mathbf{q}, \omega) = \frac{n_2}{\hbar} \left(\frac{1}{\omega - \dfrac{\hbar q^2}{2m_2} + iO^+} - \frac{1}{\omega + \dfrac{\hbar q^2}{2m_2} + iO^+} \right). \tag{49}$$

Since $\epsilon_{11}(\mathbf{q}, \omega)$ is known (Vashishta and Singwi 1972), eqn. (47) may be solved by an iterative procedure. Then the annihilation rate can be computed from $g_{21}(0)$.

Equation (47) differs from that of Sjölander and Stott by the presence of the derivative, $a_{21}n_1\partial/\partial n_1$, the properties and relevance of which are discussed in full by Vashishta and Singwi. Bhattachrayya and Singwi deduce that $a_{21} = 0{\cdot}11$ is appropriate for the entire metallic density range and, its dramatic effect in removing the unphysical pile-up of electrons around the positron for $r_s > 5$ A.U. can be seen in fig. 10. It is obvious that this approach is in many respects superior to those discussed earlier. Nevertheless the earlier theories have been of considerable value in providing both conceptual pictures and numerical results which have acted as a spur for experiment and theory alike.

2.4.3. *Positron lifetimes in real metals*

Generalizations of the electron gas theories of the previous section, aimed at providing a better description of annihilation in real metals, have generally been guided by the ideas that emerge from a one-electron approach to the analysis of positron studies in metals. Reference to § 3.1 will suggest that the necessary first steps should include the consideration of both the periodic nature of the positron and conduction electron states and the annihilation of core electrons.

Hede and Carbotte (1972) have considered annihilation in a ' nearly free ' electron band. A modified ladder approximation includes the effects of a weak electron–lattice interaction. The periodic lattice potential introduces higher momentum components, $\hbar(\mathbf{k} + \mathbf{G})$, into the wavefunctions of electrons of wavevector $\mathbf{k}$ (Berko and Plaskett 1958). The weak lattice model adopted by Hede and Carbotte takes as zero, higher momentum components for other than nearest-neighbour reciprocal lattice vectors $\mathbf{G}$. The pertinent question is then the way in which the various momentum components are enhanced by the electron–positron interaction. As a result of their analysis the authors present the following picture. The momentum dependent annihilation rate, or angular distribution is approximately a constant factor times the uncorrelated contributions as deduced by a one-electron analysis. A precise investigation would reveal more subtle distortion of the various contributions very similar to the bulging or swelling already described for the free-electron parabola. This description of conduction electron distributions should be realistic for the monovalent alkali metals.

A rather different situation has been discussed by Fujiwara (1970) who considers states close to a zone face. Here the electron–lattice interaction is strong and Fujiwara predicts a large variation in enhancement factors across the zone face. The rather severe approximations that are required for algebraic tractability somewhat lessen the significance of the results. Most serious is the use of a plane wave positron state since as Lock *et al.* (1973) have shown, a considerable variation in even the uncorrelated annihilation rate can occur as a result of interplay between positron and electron higher momentum components.

The effect of core annihilation is easily seen in fig. 10 where the experimental rates are generally larger than the electron gas values. Core annihilation rates and enhancement factors for sodium and aluminium have been calculated in the ladder approximation (Carbotte and Salvadori 1967). A generalization (Carbotte 1966) of the electron gas theory allows the Bloch nature of the particle states to be included. The calculated core enhancement factors are almost momentum-independent and do little to change the shape of the distributions obtained from the independent particle approach (§ 3.1). The enhancement is significant but small in comparison with that for the conduction electrons, a result which probably justifies the neglect of any core electron contribution to the screening of the positron force.

Some guide to a more general interpretation of the total annihilation rate in any metal can be obtained from an elementary analysis of the angular distribution. When an angular distribution consists of several components arising from annihilation with different classes, bands, or groups of electrons it is necessary, as one can easily verify from eqns. (16), (25), that the area under each component be proportional to the partial annihilation rate with the corresponding group of electrons. In some metals, the relative area of the broad component mainly associated with core electrons is large and, if we accept the association, we may combine the results from experimental lifetime and angular correlation studies to obtain values for the partial annihilation rates by writing the total rate

$$\Gamma = \Gamma_v + \Gamma_c \quad \text{and} \quad \Gamma_v/\Gamma_c = A_v/A_c. \tag{50}$$

Γ_v and Γ_c are respectively, the valence (conduction) and core annihilation rates and A_v and A_c the areas of the corresponding parts of the angular distribution. (A discussion of the validity of the interpolation procedure by which A_v and A_c are deduced will be given in § 3.2).

A comparison of Γ_v and Γ_c with the results of one-electron calculations shows that both contributions are considerably enhanced by many-electron effects. Γ_v however is invariably smaller than the theoretical electron gas rate. Such a result is not surprising and is plausibly attributed to a core electron contribution to the screening of the positron force which thus reduces the response of the valence electrons and, in consequence, Γ_v.

These observations suggest a simple method (West 1971) for the comparison of measured rates and electron gas theory. In the absence of what would clearly be difficult calculations we recognize the value of the angular distribution in providing a measure of the relative polarizability of core and valence electrons. Reference to the discussions of the previous section suggests a similar relative contribution to the screening of the positron force. If we believe that the electron gas theory provides a reasonable guide to the general factors determining the annihilation rates in real metals, we may take account of the participation of core states in *all* aspects of the polarization by renormalizing the usual valence electron density, ρ_0, according to the prescription

$$\rho = \rho_0 \{1 + A_c/A_v\}. \tag{51}$$

The result of this renormalization, displayed in fig. 15, can be seen to cause the experimental rates to fall close to the theoretical electron gas curve. Although this analysis cannot be regarded as providing more than a semi-quantitative account of correlation effects in real metals its comparative success would seem to confirm a reasonable consistency between the pictures offered by lifetime and angular correlation studies.

Fig. 15

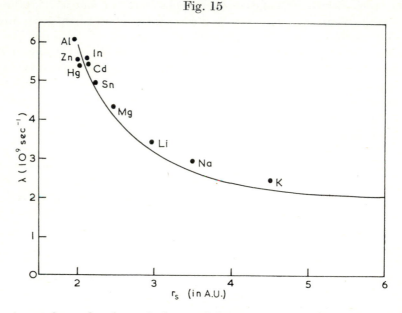

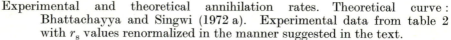

Experimental and theoretical annihilation rates. Theoretical curve : Bhattachayya and Singwi (1972 a). Experimental data from table 2 with r_s values renormalized in the manner suggested in the text.

Such a generalization of electron gas theories is clearly not applicable to semiconductors where band effects inhibit the electron polarization so as to cause the experimental annihilation rates to fall well below the electron gas curves. A strong correlation between band gap, electron polarizability and thus annihilation rate, in diamond, germanium, and silicon has been exposed by Fieschi *et al.* (1968).

2.4.4. *Positron effective mass and thermalization*

The earliest attempts at a quantitative interpretation of the angular distributions for metals (Stewart 1957) were often confined to the already discussed extraction of a Fermi momentum, k_F, from the position at which a break or discontinuity is seen in the curve. Majumdar (1965 a) has suggested that, notwithstanding correlation effects, a sharp break will always occur at the unperturbed value of k_F. In a real experiment at finite temperatures various effects will tend to smear this break. Aside from the unavoidable but minimizable effect of finite angular resolution the most important contribution to this smearing is provided by the thermal motion of the positron.

Since the annihilation rate is almost velocity independent the momentum distribution, for a low concentration of positrons, can be expected to approximate to a Boltzmann distribution appropriate to the sample temperature T :

$$F_+(p) = [1/(2\pi m_+ {}^* k_B T)^{3/2}] \exp [-p^2/(2m_+ {}^* k_B T)]. \qquad (52)$$

Here the symbols have their usual meaning and the inclusion of an effective mass, $m_+{}^*$, rather than the bare mass, m_0, anticipates the possibility that the motion of the positron is significantly affected by interaction with various excitations of the surrounding medium.

The effect, on the angular distribution, of the thermal smearing of the Fermi–Dirac-electron distribution is much less important than the positron effect (Stewart 1967) and the electron momentum distribution for a polycrystalline simple metal may be reasonably taken as the theta function, $\theta(k_F - k)$. Thus the long slit distribution,

$$N(p_2)\, dp_z \propto \int d^3\mathbf{k}\, dp_x\, dp_y\; \theta(k_F - k) F_+(|\mathbf{p} - \mathbf{k}|)\, dp_z. \qquad (53)$$

The integration of this equation is straightforward and a comparison with experimental distributions leads directly to the measurement of an effective mass (Stewart and Shand 1966, Majumdar 1965 b, Brandt *et al.* 1966). Kim *et al.* (1967) have measured the angular distributions in alkali metals over a wide range of temperatures and deduce, for example, a positron effective mass in sodium, $m_+{}^* = (1 \cdot 8 \pm 0 \cdot 2)m_0$.

There are several reasons why we might expect the positron effective mass to be different from the bare mass. In an ordered solid, the positron energy band structure provides one contribution. Majumdar (1966) has computed a band effective mass $\sim 1 \cdot 06 m_0$ for positrons in sodium. Further contributions which result from positron interactions with both electrons and phonons have been assessed by propagator techniques similar to those employed in the calculations of positron lifetimes. An effective mass approximation to the positron energy can be obtained from the results of a perturbation calculation of the self-energy part (Mattuck 1967) of the single-particle positron Green's function. Electron gas calculations of the effective mass due to positron–electron correlation in alkali metals (Hamann 1966, Bergersen and Pajanne 1969) provide values of the order of $1 \cdot 1 m_0$ which even when combined with the band result still fall short of the measured values. However similar calculations of the effect of positron–phonon interactions (Mikeska 1967, Mikeska 1970, Bergersen and Pajanne 1971) predict an equilibrium positron momentum

distribution which deviates significantly from Maxwellian form, in such a
way as to provide rather more smearing than that to be expected (in terms
of eqns. (52), (53)) from a relatively small additional contribution to the
effective mass.

A definitive statement about the agreement between theory and experiment is as yet not possible since the measured masses can be extracted
from the experimental distributions in a variety of ways (see, for example,
Brandt *et al*. 1966, Stewart and Shand 1966), and the accuracy of the data
is not sufficient to resolve the precise form of the smearing function.
Furthermore, the relevance of the comparison can be questioned (Arponen
1971). The theoretical masses are derived from the properties of a one-
particle Green's function whereas the angular distributions reflect the
properties of the electron–positron two-particle function. We have
already noted in § 2.4.2 that self-energy corrections to the one-particle
propagators may not be strongly reflected in the pair momentum distribution. Nevertheless the results obtained by Kim *et al*. (1967), shown in
fig. 16, are clearly consistent with the effective mass picture represented
by eqns. (52), (53).

Fig. 16

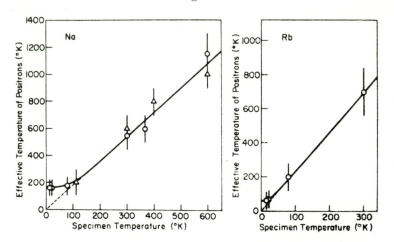

Positron effective temperatures in Na and Rb. The experimental points have
been obtained from an analysis of smearing of angular distributions.
The analysis assumes a positron momentum distribution of the form
$\exp\left(-p^2/2mk_{\mathrm{B}}T\right)$. Thus the slopes of the graphs also define the effective
mass ratio m_+^*/m_{e} (Kim *et al*. 1967).

If the linear portions of these curves represent the temperature-
controlled equilibrium behaviour of the thermalized positrons their non-
linear form at low temperatures would seem to indicate that some positrons
are not thermalized at temperatures below 100°K. (More sophisticated
interpretations are possible (Arponen 1970).) The thermalization of

positrons by excitation of electrons and phonons may also be studied through the properties of the positron Green's function. Calculations of the times required for thermalization through excitation of electron–hole pairs (Carbotte and Arora 1967) give values somewhat larger than those deduced by Lee-Whiting (1955), the discrepancy being attributable to the somewhat arbitrary choice of positron–electron interaction made in the earlier work. The thermalization times for a temperature of 300°K are now (Carbotte and Arora 1967) an appreciable fraction of measured life-times and are little affected by the detailed choice of interaction. In some metals at low temperatures the predicted thermalization times are longer than the lifetime.

The fullest description of the experimental results of fig. 16, which takes account of the essentially statistical nature of the thermalization process, is provided by the Boltzmann equation approach used by Woll and Carbotte (1967). An initial positron momentum distribution decays by depletion through annihilation and scattering with conduction electrons via a suitable screened interaction. The Boltzmann equation is solved numerically for the final stages of thermalization using a variety of starting distributions, all of which are found to relax eventually to the same Maxwellian distribution. At sufficiently low temperatures, annihilation occurs from a non-equilibrium situation, and a non-linear dependence of the form shown in fig. 16 results. The calculations assume the experimental values of effective mass and while they reproduce the general form of the experimental results there remains a systematic disagreement between the experimental and calculated values of minimum positron effective temperature.

The previously mentioned work of Mikeska (1970) suggests that the positron–phonon interaction, consistent with its contribution to an effective mass, also plays a significant role in the thermalization process. However, although the additional scattering tends to shorten the thermalization time the consequent distortion of the positron momentum distribution may increase the thermal smearing of the angular correlation curve. Again a precise comparison between experiment and theory is ruled out by practical and theoretical problems.

A preoccupation with the properties of the positron Green's function leads almost inevitably to the question of the possible existence of bound positronium-like states. Such a possibility is also suggested by the tendency of positron lifetimes to approach the spin-averaged positronium value as the density is reduced (fig. 10). The breakdown of various approximate methods of lifetime calculation (Crowell *et al.* 1966, Sjölander and Stott 1970) has also been identified with bound state formation although the most recent calculations (Bhattacharyya and Singwi 1972 a) suggest that the breakdown results from an inadequate theory rather than an underlying physical process. The possibility of a bound state at sufficiently low densities has been considered by many workers (Callaway 1959, Held and Kahana 1964, Kanazawa *et al.* 1965, Bergersen and Terrell

1968, Arponen and Jauho 1968, Bergersen 1969, Majumdar and Rajagopal 1970, Arponen 1970). The subject is somewhat controversial, but from an experimental viewpoint, largely academic, since all agree that the bound state cannot exist at metallic densities. The possibility of positron bound states with more massive partners is more credible and, as we shall see in chapter 4, suggested by many experiments.

3. ELECTRONIC STRUCTURE STUDIES

3.1. *The independent particle approach*

3.1.1. *Positron wavefunctions*

The relevance of angular correlation studies to the electronic structure of some materials was clearly demonstrated by the early experiments of Stewart (1957) and De Benedetti *et al.* (1950). Initial attempts to explain the results of these and similar experiments made use of the simplest approximate electron wavefunctions and crude model positron wave-functions applied to eqn. (15) (De Benedetti *et al.* 1950, Ferrell 1956). More serious attempts to provide a realistic theory based on uncorrelated wavefunctions were made by Berko and Plaskett (1958) whose analyses have provided the basis for many subsequent experimental and theoretical investigations.

The plausibility of such an approach is plainly suggested by the many-electron studies outlined in the previous chapter and its utility, in situations where a complex structure precludes a full many-electron treatment, is clear. The choice of appropriate electron wavefunctions in eqn. (15) can obviously be guided by their value in other problems involving the material of interest. In the case of a positron wavefunction a more detailed discussion seems called for.

In most materials the ground state of the positron will be at the bottom of a 1s band ($\mathbf{k} = 0$). Early calculations (Donovan and March 1958, Berko and Plaskett 1958), guided by analogy with the corresponding electron problem, made use of the Wigner–Seitz approximation of a spherical cell. Inside the cell $\psi_+(r) = R_+(r)/r$ where, in atomic units, $e = \hbar = m = 1$, $c = \alpha^{-1}$,

$$d^2 R_+(r)/dr^2 + 2(E + V(r))R_+(r) = 0. \tag{54}$$

In the spirit of the one-electron approximation $V(r)$ may be taken as the sum of the repulsive Coulomb potential of the nucleus and the potential arising from the electronic charge distribution obtained from a Hartree–Fock self-consistent calculation (Herman and Skillman 1963). A positron potential is best calculated from the tabulated electron wavefunctions since the similarly tabulated potentials frequently involve an unwanted electron exchange term (West *et al.* 1967). The radial wave eqn. (54) is numerically integrated (Hartree 1952), and the eigenvalue chosen so as to satisfy the boundary condition $d\psi_+(r)/dr = 0$ at $r = r_s$ the radius of the Wigner–Seitz cell.

Fig. 17

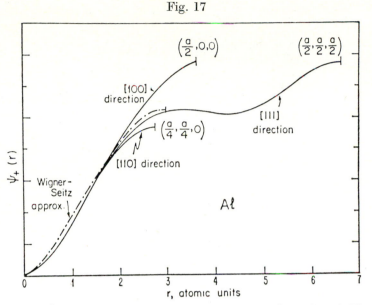

Positron wavefunctions along symmetry directions (Stroud and Ehrenreich 1968).

A typical positron wavefunction is shown in fig. 17. This wavefunction closely approaches zero at the origin, a result which is common for heavy atoms although in light atoms the amplitude, $\psi_{+}(0)$, may be significant. The amplitude at small r values will determine the annihilation rate with tightly bound electrons. Although this part of the wavefunction is probably a good representation of the positron behaviour close to the nuclei the solution is only defined for $r < r_s$ and if the occupied states extend appreciably outside this region some way of extending the definition is required. Some extensions of the Wigner–Seitz solution that have been employed (Berko and Plaskett 1958, Rose and de Benedetti 1965, West *et al.* 1967) are not satisfactory, in that the approximations are made in the very regions where the positron wavefunction is large and likely to provide a large contribution to the integral in eqn. (15). A better solution (Stroud and Ehrenreich 1968) makes use of a plane wave expansion of the positron wavefunction in terms of vectors of the reciprocal lattice. In contrast to the corresponding electron problem a fast convergence of the expansion is made possible by the absence of orthogonalization terms and the small amplitude of the positron wavefunction in regions where the potential is varying rapidly.

The field seen by the positron is taken to be due to a linear superposition of potentials $u(\mathbf{r})$ about each atomic site. Substitution of

$$\psi_{+}(\mathbf{r}) = \sum_{\mathbf{k}} A_{\mathbf{K}} \exp(i\mathbf{K} \cdot \mathbf{r}),$$

where the **K** are reciprocal lattice vectors, into the positron Schrödinger equation yields the secular equation

$$\sum_{\mathbf{k}'} [(K^2 - E)\delta_{\mathbf{K} \mathbf{K}'} + V_{\mathbf{K}-\mathbf{K}'}]A_{\mathbf{K}'} = 0. \tag{55}$$

The Fourier coefficients, $V_{\mathbf{K}}$, of the periodic potential may be deduced from X-ray scattering data (Stroud and Ehrenreich 1968) or from the Fourier transform of a single-centre potential (Gould *et al.* 1972) by means of the relation

$$V_{\mathbf{K}} = Nu(\mathbf{K}) = N \int u(\mathbf{r}) \exp(-i\mathbf{K} \cdot \mathbf{r}) \, d^3\mathbf{r} \tag{56}$$

The high symmetry at $\mathbf{k} = 0$ allows a group theoretical reduction of eqn. (55), resulting in a modestly sized secular determinant which nevertheless allows for the inclusion of many plane waves. An example of the resulting $\psi_+(\mathbf{r})$ is shown in fig. 17 and demonstrates a considerable anisotropy in interstitial regions. The eigenvalues deduced by this method are similar to those obtained in the Wigner–Seitz approximation and, in metals, usually provide a ground-state energy lying well above the periodic potential minima.

3.1.2. *The photon pair momentum distribution*

It is convenient to consider a periodic system of volume V containing N primitive cells each of volume Ω. The thermalized positron is taken to exist in a ground state, $N^{-1/2}\psi_+(\mathbf{r})$, with Bloch wave vector zero. The electron states are represented by doubly-occupied Bloch functions $\psi_{\mathbf{k}}^l(\mathbf{r}) = N^{-1/2}u_{\mathbf{k}}^l(\mathbf{r}) \exp(i\mathbf{k} \cdot \mathbf{r})$, where l is a band index and $\mathbf{k}$ is the reduced wave vector. ($u_{\mathbf{k}}^l(\mathbf{r})$ and $\psi_+(\mathbf{r})$ are normalized to one in a unit cell.)

Inserting these expressions into eqn. (15) and summing over the occupied states, we obtain, again in atomic units,

$$\Gamma(\mathbf{p}) = \frac{\alpha^3}{4\pi^2 N^2} \sum_{\substack{\text{occ.} \\ \mathbf{k}, l}} \left| \int_V d^3\mathbf{r} \exp[-i(\mathbf{p}-\mathbf{k}) \cdot \mathbf{r}]u_{\mathbf{k}}^l(\mathbf{r})\psi_+(\mathbf{r}) \right|^2. \tag{57}$$

Since both $u_{\mathbf{k}}^l(\mathbf{r})$ and $\psi_+(\mathbf{r})$ have the periodicity of the lattice,

$$\Gamma(\mathbf{p}) = \frac{\alpha^3}{4\pi^2 N^2} \sum_{\substack{\text{occ.} \\ \mathbf{k}, l}} \left| \sum_{\mathbf{R}_i} \exp[-i(\mathbf{p}-\mathbf{k}) \cdot \mathbf{R}_i \int_\Omega d^3\mathbf{r} \right.$$

$$\left. \times \exp[-i(\mathbf{p}-\mathbf{k})\mathbf{r}]u_{\mathbf{k}}^l(\mathbf{r})\psi_+(\mathbf{r}) \right|^2. \tag{58}$$

The sum over lattice vectors $\mathbf{R}_i$ may be evaluated in the usual way (Ziman 1965) whence

$$\Gamma(\mathbf{p}) = \frac{\alpha^3}{4\pi^2} \sum_{\substack{\text{occ.} \\ \mathbf{k}, l}} \sum_{\mathbf{G}} \delta_{\mathbf{p}-\mathbf{k}, \, \mathbf{G}} \, |I_\Omega^l(\mathbf{p}-\mathbf{k})|^2. \tag{59}$$

Here, **G** is any reciprocal lattice vector and

$$I_\Omega{}^l(\mathbf{p}-\mathbf{k}) = \int_\Omega d^3r \exp\left[-i(\mathbf{p}-\mathbf{k}) \cdot \mathbf{r}\right]u_\mathbf{k}{}^l(\mathbf{r})\psi_+(\mathbf{r}). \tag{60}$$

Contributions to $\Gamma'(\mathbf{p})$ from individual terms in the sum over **G** are thus spatially separated, and for each band, l, at the most one term will contribute at a particular **p**. The relative intensity of the different contributions will depend on the nature of the electron state $u_\mathbf{k}{}^l(\mathbf{r})$.

In many materials the electron states of interest may be sufficiently close to those of the free atoms to be well represented by a simple Bloch sum of atomic orbitals

$$u_\mathbf{k}{}^l(\mathbf{r}) \exp(i\mathbf{k} \cdot \mathbf{r}) = \sum_{\mathbf{R}_i} \phi(\mathbf{r}-\mathbf{R}_i) \exp(i\mathbf{k} \cdot \mathbf{R}_i).$$

With a full Brillouin zone, the sum over occupied states is easily performed and the contribution to the annihilation rate from the band labelled $n, l,$ is (Berko and Plaskett 1958)

$$\Gamma^{nl}(\mathbf{p}) \, d^3\mathbf{p} = \frac{\alpha^3}{4\pi^2} \sum_{m=-l}^{+l} \left| \int_V d^3r \right.$$

$$\left. \times \exp(-i\mathbf{p} \cdot \mathbf{r})\phi_{nlm}(\mathbf{r})\psi_+(\mathbf{r}) \right|^2 d^3\mathbf{p}. \tag{61}$$

This expression, which is also that pertaining to the extreme case of tight-binding in which there is no coupling between states on different sites, can be easily reduced to a form amenable to numerical calculation. The exponential factor and the electron orbital are expanded in appropriate radial and spherical harmonic functions. The orthogonality properties of the harmonic functions allows the m summation to be performed giving a spherically symmetric function

$$\Gamma^{nl}(\mathbf{p}) = \frac{\alpha^3(2l+1)}{\pi} \left| \int \psi_+(r)P_{nl}(r)j_l(pr)r \, dr \right|^2. \tag{62}$$

$j_l(pr)$ is a spherical Bessel function and $(1/r)P_{nl}(r)$ is the radial part of the electron wavefunction. Equation (62) may be evaluated by numerical integration and the long slit angular distribution obtained by a similar integration of eqn. (26).

Berko and Plaskett (1958) were first to use this approach with Wigner–Seitz positron wavefunctions in studies of core annihilation in aluminium and copper. The intensity of the broad component, $(p_z > p_f)$ in aluminium is small and thus the ability of their calculations to reproduce its general form cannot be regarded as very significant. Their poor reproduction of the large angle parts of the copper angular distribution may be attributed to the relatively crude description of the outermost 3d ' core ' electrons afforded by the simple tight-binding approximation. Nevertheless similar calculations for metals in which the tight-binding approximation should be more realistic are unable to reproduce the shape of experimental curves using the simple Wigner–Seitz wavefunction (Rockmore and Stewart 1967, West *et al.* 1967).

More successful is the approach of Rose and de Benedetti (1965) who adopt a ' muffin-tin ' extension of the Wigner–Seitz model to compute a momentum distribution for the 3p electrons in argon. Unfortunately, these calculations require the evaluation of awkward two-centre integrals, a complication which does not arise if one employs the plane wave expansion positron wavefunction (eqn. (55)). Gould *et al.* (1972) have used this latter method in a tight-binding treatment of core annihilation in several f.c.c. and h.c.p. metals. The results for copper are again poor and point to the predictable need for a more appropriate description of the electron states. For simpler metals, the form of the angular distribution is reproduced well over a significant range of momenta and the remaining small discrepancies can be reasonably assigned to more subtle features of the electronic structure. The results for Zn shown in fig. 18 may be taken as representative.

Fig. 18

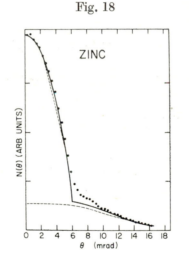

Long slit angular distribution for polycrystalline zinc. Experimental points are from Mogensen and Trumpy (1969). The solid theoretical line includes enhancement effects on the conduction electron distribution (Gould *et al.* 1972).

In all calculations of positron wavefunctions the deviation of the wavefunction from constancy, $\sum_{K \neq 0} |A_K|^2$, is small. Provided that the overlap of states on different atoms is not too great, the pair momentum distribution for a full band, when of a significant intensity, is then similar to the momentum density of the corresponding states of the free atoms. The modifications to this simple result for polyatomic systems containing non-primitive lattice translations have been considered by Brandt *et al.* (1966).

In the case of extreme tight binding of the electron states to individual atoms, $I_\Omega(\mathbf{p}-\mathbf{k}) = I_\Omega(\mathbf{p})$, as in eqn. (61).   In a cell containing two atoms

$$|I_\Omega(\mathbf{p})|^2 = |I_\Omega^{(1)}(\mathbf{p})|^2 + |I_\Omega^{(2)}(\mathbf{p})|^2$$

where $I_\Omega^{(i)}(\mathbf{p})$ is the Fourier transform of the wavefunction product in the *i*th atomic cell.   In a more general case, coupling between the states on atoms (1) and (2) will produce interference terms in the momentum distribution.   If, for example, $u_\mathbf{k}(\mathbf{r}) = u_\mathbf{k}^{(1)}(\mathbf{r}) + u_\mathbf{k}^{(2)}(\mathbf{r}) \exp(i\mathbf{k} . \mathbf{d}_{12})$, where $\mathbf{d}_{12}$ is the non-primitive vector joining atoms (1) and (2),

$$
\begin{aligned}
|I_\Omega(\mathbf{p}-\mathbf{k})|^2 = |I_\Omega^{(1)}(\mathbf{p}-\mathbf{k})|^2 + |I_\Omega^{(2)}(\mathbf{p}-\mathbf{k})|^2 \\
+ 2|I_\Omega^{(1)}(\mathbf{p}-\mathbf{k})| \, |I_\Omega^{(2)}(\mathbf{p}-\mathbf{k})| \cos(\mathbf{p}-\mathbf{k}) . \mathbf{d}_{12}. \quad (63)
\end{aligned}
$$

In some cases the additional interference term may strongly affect the form of the momentum distribution, an extreme example being observed in the angular distribution from lithium hydride (fig. 19).   In more complex systems containing non-primitive lattice vectors, the presence of such

Fig. 19

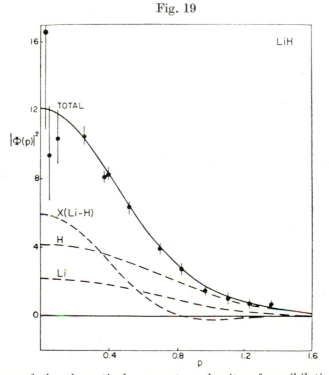

Comparison of the theoretical momentum density of annihilating positron–electron pairs in LiH with the experimental results of Stewart and March (1961).   The broken curves (Li and H) derive from the separate atomic cells and X from the interference term (Brandt *et al.* 1966).

interference terms will depend on the occupancy of bonding and anti-bonding configurations, and the final form of the momentum distribution can only be deduced from sophisticated band-structure calculations, which of course, will also display the equally important interference effects which result from the primitive periodicity of the system.

Where such effects are important it is often convenient to expose the periodicity of $u^l_{\mathbf{k}}(r)$ and $\psi_+(\mathbf{r})$ more directly by representing them both as Fourier series in reciprocal lattice vectors, $\mathbf{K}$, $\mathbf{G}$, i.e.

$$\psi_+(\mathbf{r}) = \sum_{\mathbf{K}} A_{\mathbf{K}} \exp (i\mathbf{K} \cdot \mathbf{r}), \quad u_{\mathbf{k}}{}^l(\mathbf{r}) = \sum_{\mathbf{G}} B_{\mathbf{G}}{}^l(\mathbf{k}) \exp (i\mathbf{G} \cdot \mathbf{r}). \qquad (64)$$

In these terms

$$\Gamma(\mathbf{p}) = -\frac{\alpha^3}{4\pi^2} \sum_{\substack{\text{occ.} \\ \mathbf{k}, l}} \sum_{\mathbf{G}} \delta_{\mathbf{p} - \mathbf{k} \cdot \mathbf{G}} \left| \sum_{\mathbf{K}} A_{\mathbf{K}} B_{\mathbf{G} - \mathbf{K}}{}^l(\mathbf{k}) \right|^2. \qquad (65)$$

This representation is particularly appropriate for the analysis of single-crystal studies, directed at the measurement of Fermi surface parameters. For the present we restrict our discussion to the case in which the particle states are close to plane waves. The principal part of $\Gamma(\mathbf{p})$ is then that for which $\mathbf{p} = \mathbf{k}$, and is wholly confined to a region of p-space which is an image of the occupied region of k-space, that is, the Fermi surface. The remaining contributions arise from the deviation of the particle wavefunctions from simple plane waves. These higher momentum components (HMC) produce further contributions which may extend to considerably higher momenta.

The effect of positron wavefunction HMC is shown in fig. 18. The combined effects of positron and electron HMC can be seen in calculations of the conduction electron angular distribution for lithium (fig. 20) where an analysis in terms of conduction electron structure is made more appropriate by a relatively small contribution from core annihlation (Melngailis and de Benedetti 1966). The exclusion of the positron wavefunction from regions close to the nuclei allows for a reasonable assessment of electron HMC from the components present in the smooth parts of pseudo-wavefunctions.

In materials where significant contributions arise from both nearly-free and tightly-bound electron states, a one-electron analysis must be supplemented by arbitrary enhancement factors which allow for the different polarizabilities of these various types of state. Thus the component parts of an angular distribution such as that shown in fig. 18 usually include arbitrary normalization to obtain the best fit to the experimental curve.

There are, of course, as many single-particle approximations to the theory of positron annihilation in solids as there are corresponding approximations to the electron states. Nevertheless, the two simple extremes discussed in previous paragraphs contain the essential ingredients for the description of many angular distributions, and provide a satisfactory basis for the discussion of results from more complex systems.

Fig. 20

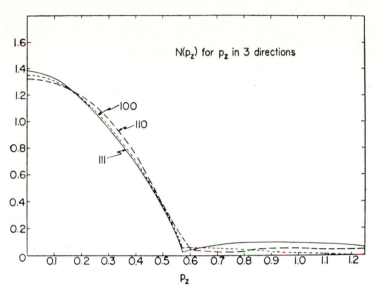

Calculated conduction electron angular distributions for lithium (Melngailis and de Benedetti 1966).

The theory presented in this section is formally similar to that of the energy line profile which results from the Compton scattering of mono-energetic photons in matter. This subject has been recently reviewed by Cooper (1971).

Subject to some reasonably well understood approximations, the line-shape of scattered photons or Compton profile, is a projection of the momentum distribution of the scattered electrons, i.e.

$$I(p_z) \propto \int \int \rho(\mathbf{p})\, dp_x\, dp_y,$$

where

$$\rho(\mathbf{p}) = \sum_n \left| \int \psi_n(\mathbf{r}) \exp\left(-i\mathbf{p} \cdot \mathbf{r}/\hbar\right) d^3\mathbf{r} \right|^2.$$

Thus, if $\psi_+(\mathbf{x}) = 1$ in eqn. (15), the Compton profile is identical to the long slit angular distribution (eqn. (25)). The absence of purely positron effects seems to offer an immediate advantage in Compton scattering studies. Inevitably other problems arise.

This simple description of a Compton profile breaks down unless the energy shift in the scattered photon is significantly larger than the binding energy of the scattering electron. Consequently experiments employing conventional X-ray sources have generally been confined to the lighter element materials. Such experiments are technically more difficult than angular correlation studies.

However, this picture should be radically altered by the arrival of solid state detectors (§ 1.3.3). The utilization of high energy monochromatic gamma-ray sources will allow an extension to heavier elements and should ease the technical problems.

Comparative positron and Compton scattering experiments are likely to be of invaluable assistance in resolving some of the interpretational problems that presently beset both techniques.

3.2. *Angular distributions and electron states in solids*

An angular distribution will provide the best reflection of the electronic structure of the containing medium when the positron annihilates form a delocalized state which permeates most regions of the solid. Thus in molecular materials, in which an appreciable fraction of positrons form positronium, an analysis of the angular distribution may be restricted to the measurement of the intensity and width of the narrow positronium

Fig. 21

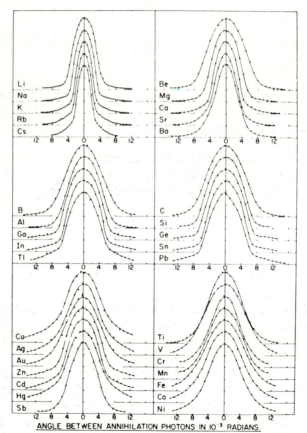

Reproduced by permission of the National Research Council of Canada.

Long slit angular distributions for 34 metals (Stewart 1957).

peak, the former providing a measure of the positronium fraction and the latter allowing some assessment of the interaction of the parapositronium atoms with the surrounding medium (Tao 1972). However, in the simpler molecular materials, an analysis of the broader component can give some indication of the states of the electrons contributing. Chuang and Hogg (1967) and Chuang *et al.* (1968) have shown that the broad components for methane, hexane, and decane are consistent with annihilation of the electrons participating in the covalent bonding. Holt *et al.* (1968) have obtained a similar result for ammonia. These results may be fortuitous since no differentiation is made between the pick-off annihilation of orthopositronium and the annihilation of unbound positrons. The writer is not aware of any similar analyses of angular distributions for more complex molecular materials.

In the case of ordered ionic solids of sufficient purity to allow but a small probability of positronium formation, annihilations from a delocalized positron state should provide a strong reflection of the momentum density of the annihilating electrons in the angular distributions. The theoretical description of the angular distribution of lithium hydride (Brandt *et al.* 1966) supports this belief. The theoretical intensities of the component parts of this distribution (fig. 19) demonstrate a plausible tendency for the positron wavefunction to favour the outer shells of negative ions. Such an effect is also apparent in the angular distributions of alkali halides (Stewart and Pope 1960). More recent investigations have been chiefly directed at the elucidation of the problems of trapping by defects and the great variety of traps and positron states suggested by these studies (§ 4.2.1) has justifiably inhibited further attempts at detailed analysis in terms of electronic structure.

The angular distributions for metals and semiconductors provide better material for analysis and some attempt at classification is made possible by the extensive results of Stewart (1957), reproduced in fig. 21.

The major part of the angular distributions for polycrystalline alkali metals has the parabolic form predicted by the Sommerfeld model (§ 2.4.1). The small intensity of the broader (HMC + core) component is readily explained by the compact ion cores and nearly-free conduction electron states in these metals.

In sharp contrast, the more extended electron states of the closed d-shells of the noble metals result in a large ' core ' contribution which blends smoothly into a somewhat smaller contribution from 'conduction ' electron annihilation. The distributions for transition metals are similar. In neither case can one make a distinction between the effects of core and conduction electrons, a result which nevertheless may be regarded as a confirmation of the accepted picture of electronic structure.

The angular distributions for most other metals are less easily classified. Whether or not a discontinuity between the parabolic and broad parts of an angular distribution can be discerned will plainly depend on the form of the relevant electron states. In those cases where a significant tail can be clearly resolved from the central component there seems little doubt that the predominant cause is the annihilation of core electrons (West *et al.* 1967, Arias-Limonta and Varlashkin 1970 a, Gould *et al.* 1972).

The relative intensity of a broad component will depend, very largely, on the extent to which the positron is able (or forced) to penetrate the ion cores. Crude but effective parameters are the valence and relative volume of ion cores (fig. 22). Thus in the first (Li, Be) and second (Na, Mg, Al, Si) short periods the compensating effects of increasing valence and relative ion volume cause core distributions of small and similar intensity. In the periodic sequences (Cu, Zn, Ga, Ge), (Ag, Cd, In, Sn, Sb) and (Au, Hg, Tl, Pb) the combined effects of volume and charge result in a progressive decrease of the core intensity.

Fig. 22

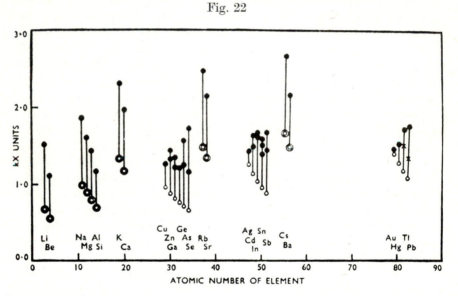

Interatomic distance and ionic radii. ● : ½ interatomic distance in metal. ⊙ : Zachariasen univalent ionic radii. ○ : Pauling univalent ionic radii. × : Goldschmidt empirical ionic radii. (Hume-Rothery and Raynor 1938).

The remaining part of the angular distribution for many materials has the parabolic form (eqn. (27)) suggested by the Sommerfeld model. Its width is determined by the electron density for states predominantly of a plane wave nature. However, the evidence of the irrelevance of parabolic form to classic metallic properties is clearly supplied by the distributions for semi-metals like bismuth, and the semi-conductors germanium and silicon. In such cases the ' parabolic ' nature of the curve results from an inadequate resolution of the comparatively small differences between the integration (eqn. (25)) of a spherical Fermi distribution and the spherical average of a similar integration of filled Brillouin or Jones zones (Jones 1960).

The measurement of conduction electron density or ' valence ' through an assessment of the width of angular distributions has been the basis of various investigations. Burton and Jura (1968) have measured the angular distribution for aluminium at pressures $0 \rightarrow 100$ kbar using a Bridgmann anvil technique. Over this range of pressures the width of the conduction electron distribution is consistent with the predictions of the free-electron theory and a valence of three.

The assessment of ' valence ' from the angular distributions of materials having more complex electronic structure is less exact but may be pursued. As a result of a comparison of free-electron parabolas with the low-angle parts of experimental distributions for certain rare-earth elements Gustafson and Mackintosh (1964) deduce a ' valence ' of two for ytterbium and three for cerium and gadolinium. Gustafson and colleagues (Gustafson *et al.* 1969, Gempel *et al.* 1972) have performed similar measurements on high pressure phases of cerium. Marked changes in electric and magnetic properties on compression of cerium have been attributed to various modifications in electronic structure, one example being the ' promotion ' of 4f electrons to the conduction band. Angular distributions for the γ, α and α' phases have similar widths and are consistent with a common valence of the order of three free electrons per atom. These authors conclude that despite the inability of the angular distributions to resolve the underlying electronic changes their results are not consistent with the promotional model.

Chouinard *et al.* (1969) have studied the angular correlation in cerium hydride for hydrogen concentrations in the range $H/Cl = 1 \cdot 8 \rightarrow 2 \cdot 8$. The increasing width of the angular distributions with increasing hydrogen concentration is interpreted in terms of a metallic model in which the hydrogen atoms donate their electrons to a conduction band. Similar effects in the systems titanium–hydrogen (Weselowski *et al.* 1963) and palladium–hydrogen (Dekhtyar and Shevchenko 1972) have also been explained by partial or complete ionization of hydrogen atoms. An alternative anionic or hydridic model assumes that hydrogen atoms capture electrons from the conduction band and is used by Green *et al.* (1971) in the interpretation of their lifetime measurements in lanthanum hydrides. We do not find these authors' arguments very convincing but angular correlation studies of this system would be appropriate.

The question of the nature of the electrons donated by the metal atoms in alkali metal–ammonia solutions has prompted a series of investigations. The observed electrical and thermal conductivities of these solutions suggest that the solvated electrons are, in some sense, almost as free as the conduction electrons of a metal. Early measurements for Li–, Na–, K–, Rb–, and Ca–ammonia solutions (Varlashkin and Stewart 1966, Varlashkin 1968), of maximum concentration 22 mole % (lithium–ammonia), showed almost no concentration dependence in the angular distributions which were of a form unlike that of solvent or solute. Although various models for the underlying electron and positron states have been suggested (Millett *et al.* 1967, Bhide and Majumdar 1969, Arias-Limonta and Varlashkin 1970 b), a quantitative explanation of the angular correlations or the lifetime results of Millett *et al.* (1967) is still required. Investigations of $0 \rightarrow 100$ mole % caesium in ammonia (Arias-Limonta and Varlashkin 1970 b) do display a concentration dependence in the angular distributions. The dependence is relatively smooth and no evidence of a Mott transition (Thompson 1968, Mott and Zinamon 1970) is detected.

A semi-quantitative assessment of conduction electron density is more easily made in the case of alloys of metals exhibiting the more parabolic type of angular distribution. Alloys of Se–Te provide angular distributions of a width consistent with a valence-band contribution of six electrons per atom from each component (Cangas *et al.* 1967). In alloys of metals of dissimilar valence (Stewart 1964, Badoux and Heinrich 1970) more subtle effects of impurity screening (Stern 1968) and positron ‘ trapping ’ (Kubica *et al.* 1971) may be important. We shall return to this subject in § 4.3.4.

Thus far the investigations described have been concerned with little more than a qualitative interpretation of angular distributions for polycrystalline materials. A more quantitative analysis of even the simpler type of parabolic + core distributions is made difficult by the frailty of two rather necessary assumptions :

1. The conduction (or valence) electron distribution can be distinguished.

2. Its form says something significant about the nature of the conduction (or valence) electron states.

An isolation or separation of the core contribution is an obvious prerequisite of any serious analysis of the conduction electron components. An interpolation procedure based on a core distribution of the gaussian form (Gustafson *et al.* 1963) suggested both by calculations (Berko and Plaskett 1958) and the angular distributions for inert elements (Rose and de Benedetti 1965, Arias-Limonta and Varlashkin 1970 a) is usual, but may be in error. The shape of the core distribution at large angles will depend largely on the form of the electron and positron wavefunctions deep in the ion cores and thus, at these angles, good agreement between theory and experiment (fig. 18) is not too surprising. However, the normal tight-binding calculations provide a relatively poor description of the real electron states in the interstitial regions of a solid, and the arguably more extended nature of these states will cause a heaping up of the core distribution at low angles (Melngailis 1970). Thus a sensitivity to structure or density changes should be most pronounced in the low angle regions.

MacKenzie, LeBlanc and McKee (1971) have found that the pressure dependence of positron annihilation rates in the alkali metals is somewhat greater than would be expected from density changes applied to electron

gas theories.   An explanation in terms of a pressure dependent core annihilation rate was suggested by these authors to be inconsistent with their
observations of the pressure induced changes in the Doppler-broadened
lineshapes.   However, their results are not inconsistent with a pressure
dependence of core distribution shape of the type suggested above.

Even when the core contribution is small, an angular distribution for a
polycrystalline material is still remarkably insensitive to the detailed form
of the valence electron states, as is well illustrated by the angular distribution for selenium.   Its form, although intermediate between, is equally
well approximated by the rather similar theoretical distributions for either
a free electron model or free atom Hartree–Fock orbitals (Hautojärvi and
Jauho 1967).

Analyses of the angular distributions for oriented single crystals are
more promising.   Here the exploitation of crystal symmetry allows significant information to be obtained from anisotropies and local structure in
angular distributions.

3.3. *Fermi surface studies*

3.3.1. *Background*

A long slit angular distribution for an oriented single crystal is conveniently distinguished by the labelling

$$N_{(ijk)}(p_z) = A \int_{-\infty}^{+\infty} \int_{-\infty}^{+\infty} \Gamma(\mathbf{p}) \, dp_x \, dp_y, \qquad (66)$$

where ijk are the indices of the lattice planes perpendicular to the p_z-
direction and A is a normalization constant.

The first significant measurements on single crystals were made by
Berko *et al.* (1957) whose studies with graphite crystals oriented with
hexagonal planes perpendicular and parallel to the detector slits disclosed
a marked anisotropy in qualitative agreement with a polar orbital picture.
Similar anisotropies were encountered in the angular distributions for
silicon and germanium (Colombino *et al.* 1964, Erskine and McGervey
1966).   The latter authors showed that a fair description of their angular
distributions could be obtained from a nearly-free electron approximation
to the states for a filled Jones zone (Jones 1960) containing four electrons
per atom.

Anisotropies in metal distributions are generally small although in the
case of lithium (Donaghy and Stewart 1967 a, Melngailis and de Benedetti
1966) and beryllium (Stewart *et al.* 1962, Shand 1969) sufficient to have
stimulated fairly complete theoretical descriptions.   Angular distributions for magnesium (Berko 1962), zinc (Kusmiss and Swanson 1968), tin
and bismuth (Mogensen and Trumpy 1969) display anisotropies rather too
small to justify a sophisticated analysis.

For light, comparatively simple metals, a one-electron calculation
supplemented by the many-body enhancement factors deduced from

electron gas models is not too difficult to justify. The relevance of a similar approach to the angular distributions for more complex materials was
demonstrated by Williams and Mackintosh (1968) with measurements on
a series of rare-earth metals and alloys. Angular distributions taken
perpendicular and parallel to the c-direction were qualitatively in agreement with APW calculations of electronic structure (Loucks 1966, Keeton
and Loucks 1968).

Despite the somewhat limited ability of these studies to resolve the more
subtle features of electronic structure the practical difficulty of refining
the rare-earth metals to the purity required for the more conventional
Fermi surface measurement techniques makes positron annihilation an
invaluable tool. A similar argument holds for Fermi surface investigations in alloys and it is to this end that the positron technique has been
most exploited.

The description of the photon pair momentum distribution offered by
eqn. (65) is most transparent in the case of monovalent metals and it is
not surprising that the largest number of investigations have been concerned with copper and its alloys for which a considerable amount of
information exists or can be argued from rigid-band models of alloying
(Ziman 1961). (A further attraction is the possibility of increased positron
flux with the utilization of in situ ^{64}Cu sources obtained by prior irradiation in the sample.) These investigations have prompted sophistications
in experimental technique and data analysis whose strengths and weak-
nesses are most easily discussed with copper as the example.

3.3.2. Copper and its alloys

An understanding of the precise relationship between an angular distribution and the various aspects of electronic structure, particularly that
of Fermi surface topology, can best be obtained from a further discussion
of eqn. (65) which, for later convenience, we re-write in the form

$$\Gamma(\mathbf{p}) = \text{const.} \sum_{\substack{\text{occ.} \\ \mathbf{k},\, l}} \sum_{\mathbf{G}} |C_\mathbf{G}{}^l(\mathbf{k})|^2 \delta_{\mathbf{p}-\mathbf{k},\, \mathbf{G}}. \tag{67}$$

$$C_\mathbf{G}{}^l(\mathbf{k}) = \sum_{\mathbf{K}} A_\mathbf{K} B_{\mathbf{G}-\mathbf{K}}{}^l(\mathbf{k}), \tag{68}$$

is the Fourier transform of the wavefunction product $u_\mathbf{k}{}^l(\mathbf{r})\psi_+(\mathbf{r})$.

$\Gamma(\mathbf{p})$ has the point symmetry of the reciprocal lattice (Mijnarends 1967).
The contribution to $\Gamma(\mathbf{p})$ from states in full Brillouin zones will be con-
tinuous (Berko and Plaskett 1958), whereas the contribution from a partially filled zone will be non-vanishing only at momentum values $\mathbf{p}$ such
that $\mathbf{p} - \mathbf{G}$ lies inside the occupied region. It is this latter contribution
which will reflect the Fermi surface topology.

The form of the long slit angular distribution (eqn. (56)) is most easily
deduced for a hypothetical model in which annihilation is restricted to
conduction electron states for which the $C_\mathbf{G}$ are $\mathbf{k}$-independent. Then

$N_{(ij)k}(p_z)$ is proportional to the cross-sectional areas through an array of ' images ' of the Fermi surface. The ' images ' are weighted by factors $|C_\mathbf{G}|^2$ and centred about values of $p_z = \mathbf{G} \cdot \mathbf{n}_{(ijk)}$ where $\mathbf{n}_{(ijk)}$ is the unit vector in the p_z-direction.

In a real metal this simple picture is upset by two important effects.

(i) Distortion of the Fermi surface ' images ' arising from the probably unknown $\mathbf{k}$-dependence of the $C_\mathbf{G}(\mathbf{k})$ for conduction electron states.

(ii) The presence of an additional, often larger, contribution from states in the filled Brillouin zones.

The first of these effects is unavoidable with any technique which probes a momentum distribution rather than the occupancy of wave-vectors $\mathbf{k}$. The complication provided by (ii) is particularly severe with the long slit geometry since, as a result of the double integration (eqn. (25)), high momentum core events are projected into small angle parts of the curve where Fermi surface structure is most likely to be observed.

Berko *et al.* (1968) attack this latter problem in copper by assuming an isotropic ' core ' distribution. Anisotropies, exhibited in difference curves such as $\{N_{(110)}(p_z) - N_{(100)}(p_z)\}$ are then attributed to conduction band annihilations and, in the range $-p_f < p_z < +p_f$, are found to be

Fig. 23

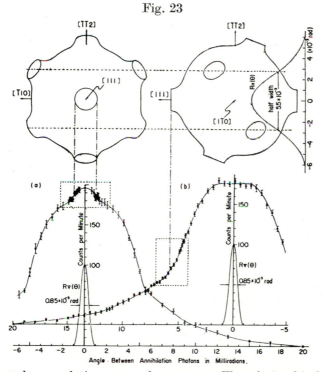

Short slit angular correlation curves for copper. The relationship between the widths of the vertical and horizontal resolution curves and Fermi surface dimensions is also shown (Fujiwara and Sueoka 1966).

roughly in agreement with predictions based on Fermi surface (Roaf 1962)
cross-sectional areas.   Anisotropies at larger angles (or momenta) can be
plausibly attributed to HMC in the conduction electron states.

The effect of the core contribution is less severe in the case of the short,
or crossed slit geometry (Fujiwara and Sueoka 1966).   In this arrangement, finite length detector slits introduce a restricted integration over p_y
(see fig. 4) defined by a resolution function of characteristic width $\sim p_f$,
the Fermi momentum.   The collimation limits the contribution from high
momentum core events and thus increases the relative structure due to
Fermi surface effects.   The resulting resolution of one particular feature
of the momentum distribution, namely that arising from states in the
necks at the $\langle 111 \rangle$ Brillouin zone faces (fig. 23), is sufficient to provide
some measure of the diameter of these necks.   However, a more general
analysis of this type of distribution is made difficult by the introduction
of a resolution function of a similar width to that of the momentum distribution.

Fig. 24

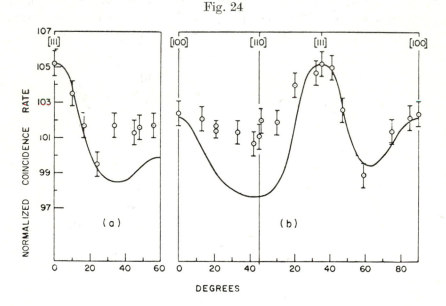

Momentum distribution anisotropies obtained from a collinear point geometry-
    rotating crystal study of Cu.   The solid line is based on a 50% isotropic
    core distribution added to an anisotropic conduction electron distribution
    deduced from the Roaf (1962) surface (Williams et al. 1968).

The logical extension of the short slit modification results in the point
detector geometry (Colombino et al. 1963) in which both p_y and p_z are
accurately determined.   Williams et al. (1968) adopt a collinear arrangement, in which the point detectors are held fixed on the p_x-axis, and the
coincidence counting rate is measured as a function of the orientation of a

rotatable single-crystal specimen. If conduction electron HMC were zero, and core annihilation absent, the quantity measured would be pro-portional to the Fermi surface diameter in the p_x-direction. Williams and his colleagues compare their copper results with a theoretical curve which assumes a 50% isotropic core contribution added to a conduction electron contribution in which the anisotropies arise from the variation of the radius vectors of the Roaf (1962) Fermi surface. The differences (fig. 24) between the experimental points and the theoretical curve are then attributed to a small ($\sim 8\%$) increase in the core contribution for orienta-tions close to the $\langle 110 \rangle$ orientation. Nevertheless, the neglect of conduc-tion electron HMC could easily account for some of the discrepancy shown in fig. 24.

The difficulty of a reliable assessment of the core contribution with the rotating crystal method has prompted more conventional point detector studies in which the detectors are moved so as to provide a distribution in two components of momentum

$$N(p_y, p_z) = A \int_{-\infty}^{+\infty} \Gamma(\mathbf{p}) \, dp_x. \tag{69}$$

This distribution obviously contains more information about $\Gamma(\mathbf{p})$ than the doubly-integrated long slit distribution. In particular, for certain orientations of a monovalent crystal, there are some regions of p_y, p_z space where there can be no contribution from electrons in states occupying the partially filled Brillouin zone, since the corresponding momentum distribution is wholly confined to regions lying within the periodically repeated Fermi surface (fig. 25). The application of this result to the experimental and theoretical point detector studies of Senicki *et al.* (1973) re-emphasizes the inadequacy of a simple tight-binding description of core annihilation in copper. Thus the theoretical deduction (Senicki *et al.* 1972) of a core anisotropy qualitatively similar to that suggested by Williams *et al.* (1968) may not be particularly significant.

Adequate angular resolution in point detector studies results in con-siderably lower counting rates than are attainable with the other geo-metrics and various multi-detector systems which might reduce this problem have been suggested (Trifthäuser 1971). Howells and Osmon (1972) have performed an experiment on copper using a spark chamber (detector 1), viewed by a pair of television cameras, and triggered by the coincident signals from a scintillation counter (detector 2). Although the poor statistics limited an analysis of the results to a demonstration of the existence of the Fermi surface necks the technological achievement re-presented by these measurements is considerable.

Experimental developments generally move towards the measurement of quantities more closely related to the total momentum distribution $\Gamma(\mathbf{p})$. In principle, the operation of a Ge(Li) detector in coincidence with a point detector scintillation counter assembly could allow the determina-tion of all momentum components of an annihilation event. At present

Fig. 25

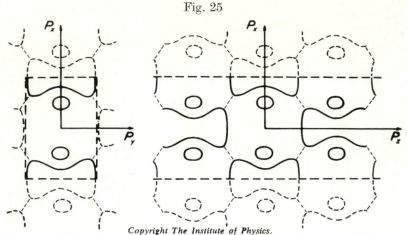

The Fermi surface of Cu in a repeated zone scheme. The coordinate axes are oriented parallel to the $\langle 110 \rangle$ directions. The broken lines depict the limits of the restricted integration in eqn. (72) (Lock *et al.* 1973).

the resolution of solid-state detectors would seem to render such a technique of marginal value.

From a theoretical point of view, the precise measurement of a sufficient number of long slit distributions for different sample orientations would provide the necessary information for the reconstruction of $\Gamma(\mathbf{p})$. Since $\Gamma(\mathbf{p})$ possesses the same point symmetry as the reciprocal lattice it may be expanded in appropriate harmonic functions. Mijnarends (1967) has described how $\Gamma(\mathbf{p})$ may then be deduced from a mathematical inversion of the consequential equations satisfied by the long slit distributions.

Application of this method (Mijnarends 1969) to eight precise long slit distributions for copper provides the momentum distribution represented by the contour diagram shown in fig. 26. Within the region bounded by the accepted Fermi surface there is a clear indication of the necks at the $\langle 111 \rangle$ zone faces. Within the necks the momentum density is lowered in a manner consistent with the nearly-free electron approximation (Berko and Plaskett 1958). Discrepancies between extracted Fermi surface radius vectors and those of Roaf (1962) or Halse (1969) are similar to the discrepancies shown in fig. 24. However, in this case the anisotropy of the momentum distribution in those regions where only states in full Brillouin zones can contribute confirms an anisotropy of the core distribution. The more recent theoretical calculations of Mijnarends (1973 a) also show this effect. Some of the discrepancy between the experimental (Mijnarends 1969) and accepted pictures of the Fermi surface topology are attributed by that author to positron–electron correlations.

Majumdar (1971) has discussed Mijnarends' method and its limitations, and has also considered the problem of deducing $\Gamma(\mathbf{p})$ from point detector results (eqn. (69)).

Fig. 26

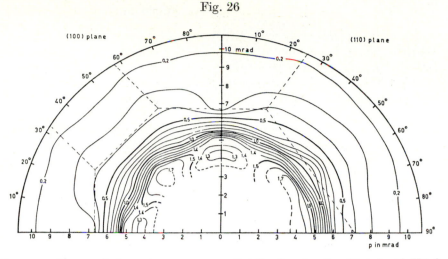

The central part of the pair momentum distribution for Cu, $\Gamma(\mathbf{p})$, in $(A.U.)^{-3}$, in the (100) and (110) planes.   (Mijnarends 1969).

The experimental and theoretical techniques described in previous paragraphs are clearly of considerable importance in the general search for information about electronic structure.   A method of analysis of long slit distributions which may provide a closer reflection of the Fermi surface has been described by Lock *et al.* (1973).

Identical long slit angular distributions are superimposed at intervals equal to that of the projections P_z of all reciprocal lattice vectors on the p_z-direction, i.e.

$$F_{(ijk)}(p_z) = \sum_{P_z} N_{(ijk)}(p_z + P_z). \tag{70}$$

From eqn. (66),

$$F_{(ijk)}(p_z) = A \sum_{P_z} \int_{-\infty}^{+\infty} \int_{-\infty}^{+\infty} \rho(\mathbf{p} + P_z \mathbf{n}_{(ijk)})\, dp_x\, dp_y. \tag{71}$$

The integration over all p_x, p_y may be replaced by a restricted integration over an appropriate primitive zone in the plane of reciprocal lattice points at P_z (fig. 25), together with a summation over those reciprocal lattice points $\mathbf{P}(P_z)$.   The resulting sum, $\sum_{P_z} \sum_{\mathbf{P}(P_z)} = \sum_{\mathbf{P}}$, when applied to a momentum distribution of the form, eqn. (67), provides the result

$$F_{(ijk)}(p_z) = \text{const.} \sum_{\mathbf{H}} \int \int dp_x\, dp_y \sum_{\substack{\text{occ.} \\ k, l}} \delta_{\mathbf{p-k.}\ \mathbf{H}} \sum_{\mathbf{G}} |C_{\mathbf{G}}{}^l(\mathbf{k})|^2. \tag{72}$$

Unlike the contributions to $N(p_z)$ which are spatially separated (see § 3.1), the contributions to $F(p_z)$ from different terms in the sum over $\mathbf{G}$ are superimposed.   Further, if positron HMC are ignored the sum,

$\sum_{G} |C_{G}(\mathbf{k})|^2$, is constant, and provides a simple multiplicative constant which may be absorbed with that already present in eqn. (72). The contribution to $F(p_z)$ from full zones is now a constant and cannot contribute to any structure. The contribution from unfilled zones is proportional to the cross-sectional areas of those parts of the Fermi surface, in a repeated zone scheme which lie within the region bounded by the broken lines in fig. 25.

The recovery of a set of undistorted images of the Fermi surface depends upon the approximation of a plane wave (constant) positron wavefunction. Consideration of less restricted conditions that $F(p_z)$ should possess the properties outlined above suggests that the most serious errors are likely to arise in the conduction electron contributions. There is some indication that electron–positron correlation effects (Fujiwara 1970) may tend to compensate for these errors but, none the less, the residual distortion of the Fermi surfaces images should be much less than that present in $\Gamma(\mathbf{p})$ or $N(p_z)$.

The result of such an analysis of a long slit distribution for copper is shown in fig. 27. The comparison with a theoretical curve deduced from the Cu 5 Fermi surface of Halse (1969) demonstrates the ability of the analysis (in this case at least) to provide an overall picture of the Fermi surface profile.

Fig. 27

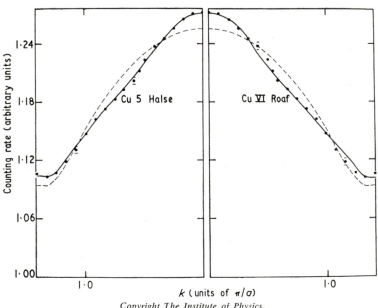

Copyright The Institute of Physics.

Fermi surface cross sectional areas perpendicular to the $\langle 110 \rangle$ directions. The experimental points are shown with curves derived from the analytic functions of Roaf (1962) and Halse (1969) as indicated, and the broken lines correspond to a spherical Fermi surface (Lock *et al.* 1973).

The same analysis provides a figure for the fraction of annihilations with electrons in states in the filled Brillouin zones. This result, when combined with plausible assumptions about the form of the corresponding contribution to $N(p_z)$ at large p_z allows one to argue the form of this ' core ' distribution for all p_z. The result is the broken line of fig. 28.

Fig. 28

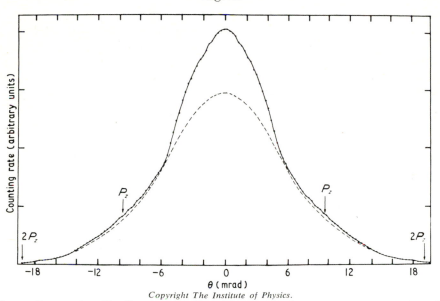

Long slit angular distribution for Cu. The resolved momentum component is parallel to a $\langle 110 \rangle$ direction of the reciprocal lattice. The broken curve encloses the contribution ascribed to annihilation from electron states in full Brillouin zones (Lock *et al.* 1973).

The magnitude of this core contribution is somewhat larger than a previous estimate (Melngailis 1970) but is consistent with estimates of the contribution to point detector results (Williams *et al.* 1968, Senicki *et al.* 1972). So large a core contribution would explain the considerable enhancement of structure in short slit studies and is in accord with suggestions made in § 3.2.

The transformation of a momentum distribution into a distribution in occupied Bloch wave vectors is more easily justified for the results of Compton scattering experiments (Cooper 1971). However, the transformation should be useful for positron studies in alloys where errors arising from positron HMC should be less dependent on composition than the Fermi surface profile. As yet, such an approach has not been tried and the interpretation of results has been largely restricted to a discussion of changes at or around the necks at the $\langle 111 \rangle$ zone faces.

We shall not dwell on the well-known problems of a Fermi surface interpretation of alloy systems for which the usual Bloch waves concepts

are complicated by potential disorder.   A discussion of the implications of
band gap or electron effective mass changes (Ziman 1961) or even the lack
of a well-defined energy–wavevector relationship (Stern 1968) can afford
to await considerably more data than is currently available.   The rigid-
band model (Ziman 1961) serves as the most convenient base for our
discussion.

Early measurements with copper–aluminium alloys containing up to
15% aluminium were made by Fujiwara et al. (1968).   Their results suggested changes in neck radii roughly in agreement with a rigid-band
prediction ; but an anomalous decrease in neck radius at 5% aluminium
concentration was also reported.   Subsequent rotating crystal—short slit
(Akahane et al. 1971), and conventional long slit studies (Murray and
McGervey 1970, Thompson et al. 1971), indicate a monotonic increase of
neck radius with aluminium concentration.   The geometrical complexity
of the rotating crystal studies complicates a quantitative analysis.   The
accumulated results of the other experiments are shown in fig. 29.

Fig. 29

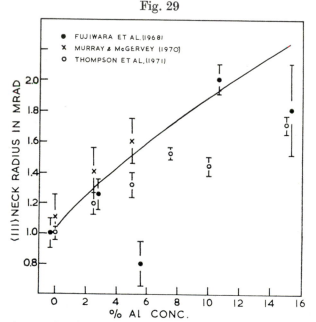

⟨111⟩ neck radius in Cu–Al alloys.   The solid line is the variation in neck radius
according to the Ziman (1961) rigid-band model (Thompson et al. 1971).

At low aluminium concentrations the variation in neck radius is consistent with the predictions of a rigid-band model.   At higher concentrations (> 7·5% aluminium) discrepancies appear.   Thompson et al. (1971)
have suggested that these discrepancies may be due to either the
appearance of short-range ordering or a tendency for positrons to favour
the copper sites.   The effects of short-range ordering in both positron

annihilation and optical properties have been discussed by Rea and De Reggi (1972). A tendency for positrons to favour the copper sites would cause ' local ' effects in some ways analogous to those observed in optical studies of density of states in alloys (Fabian 1968).

If the latter effect is important, it should be less so in the copper–nickel system, since the component elements have similar ions and similar angular distributions. The results of de Haas–van Alphen studies (Chollet and Templeton 1968) at low (< 0·1 at. %) nickel concentration are, in a rigid-band sense, consistent with each nickel atom contributing 0·4 electrons to the conduction band of the alloy.

Murray and McGervey (1970) investigated the angular distribution for an alloy of 2·3 at. % nickel and estimated, from a rigid-band analysis, a contribution of 0·8 conduction electrons from each nickel atom. Rouse and Varlashkin (1971) chose the short slit geometry for studies of alloys of 10, 50 and 90 at. % nickel and although unable to deduce a significant variation in neck radii, claimed a decrease in belly radius with increasing nickel concentration. Tanigawa and his colleagues (Tanigawa *et al.* 1971, Nanao *et al.* 1972) have also employed the short slit geometry in studies of copper–nickel alloys of 20 and 40 at. % nickel. These workers suggest that the Fermi surface is detached from the Brillouin zone boundaries in both alloys. More comparative data are clearly required but such an early detachment (at 20% nickel) is consistent with a contribution of not more than the 0·4 conduction electron per nickel atom suggested by the dH–vA analyses.

Some authors have provided more extensive interpretations for their copper alloy results. However, the reliability of some of the analyses, which depend critically on deconvolution procedures designed to remove the smearing effects of finite angular resolution, is to our mind uncertain. We have restricted our discussion in accord with this belief.

There is little doubt that positron studies are on the point of making a significant contribution to our understanding of the electronic structure of alloy systems. The recognition of positron trapping and enhancement effects (§ 4.3.4) and the exploitation of existing and possible future modifications in experimental geometry is likely to provide accelerated progress in the coming decade.

3.3.3. *Simple and not so simple metals*

We now return to a general discussion of measurements and analysis of angular distributions for single-crystal specimens.

One of the most convincing demonstrations of the value of one-electron analyses of angular distributions was given by Stroud and Ehrenreich (1968) in a study of aluminium and silicon. Fourier expansion positron wavefunctions were combined with electron wavefunctions deduced from pseudopotential calculations. The angular distribution for aluminium was almost isotropic despite anisotropies in the positron wavefunction.

The computed angular distributions for silicon were in close agreement with experimental curves obtained by Erskine as can be seen from the comparison of the large experimental and theoretical anisotropies shown in fig. 30.

Although, as was mentioned earlier (§ 3.3.1), the anisotropies in metal distributions are small, the modest intensity of core annihilation in the curves for lithium and beryllium allows for fairly sophisticated analyses in terms of conduction electron structure.

Fig. 30

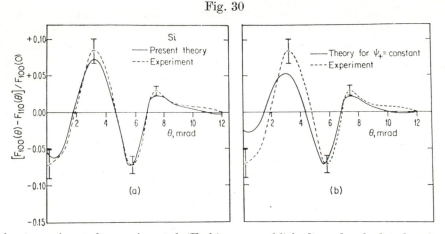

A comparison of experimental (Erskine, unpublished) and calculated anisotropies of the angular distribution for Si. (*a*) Pseudopotential theory
and Fourier expansion positron wavefunctions. (*b*) Same but with
constant positron wavefunction (Stroud and Ehrenreich 1968).

Conventional Fermi surface studies in lithium are made difficult by a phase transition at 78°K (Cracknell 1969). Donaghy and Stewart (1967 a) deduced both aspects of the Fermi surface topology and values for the most important HMC from their lithium curves. Melngailis and de Benedetti (1966) employed an OPW calculation to show that quantitative agreement with the lithium results requires consideration of Fermi surface topology, HMC, and the Kahana (1963) many-body enhancement factor. Both investigations suggested $\sim 5\%$ bulging of the Fermi surface in $\langle 110 \rangle$ directions.

Stachowiak (1970) re-interpreted the lithium data using the method of Mijnarends (1967) and suggested the possibility of necks in the $\langle 110 \rangle$ directions. However, the value of his analysis was severely limited by the small number (3) of long slit distributions available. More recently, Paciga and Williams (1971) have excluded the possibility of necks from an analysis of rotating crystal-point detector measurements. Estimates of positron and electron HMC indicate that the Fermi surface radius is $2 \cdot 9\%$ larger in the $\langle 110 \rangle$ directions than in the $\langle 100 \rangle$ and provide a value for the Fourier component V_{110} of the lattice potential $\sim 0 \cdot 10$ Ryd in good agreement with band-structure calculations by Rudge (1969).

Long slit angular distributions for beryllium (Stewart *et al.* 1962) can be compared with the results of other electronic structure studies (Watts 1964). A good description of these angular distributions was obtained by Shand (1969) from a local pseudopotential model in which Fourier coefficients of potential were chosen to fit the known Fermi surface.

The interpretation of angular distributions from more complex metals inevitably has been less successful. Williams *et al.* (1966) and Williams and Mackintosh (1968) have obtained long slit distributions for a series of rare-earth metals and an alloy. A comparison of results for yttrium with those of an APW calculation by Loucks (1966) is shown in fig. 31. The relative lack of structure in the experimental curve is attributed by these authors to electron–positron correlation effects. The angular distributions for heavy rare-earth elements are similar to those for yttrium and

Fig. 31

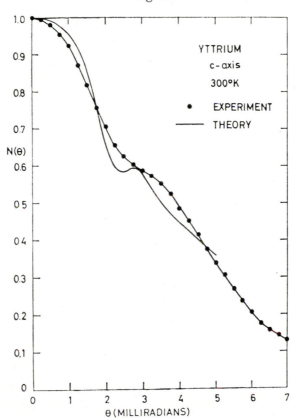

Experimental and calculated (Loucks 1966) long slit angular distribution for a Y crystal. The calculations have been corrected for experimental resolution (Williams and Mackintosh 1968).

thus in qualitative agreement with electronic structure calculations (Keeton and Loucks 1968). In holmium the effects of magnetic ordering are detected.

Kim and Buyers (1972) have measured angular distributions for molybdenum with a point detector geometry. A comparison with theoretical results from a pseudopotential calculation based on an s–d interpolation method (Ehrenreich and Hodges 1968) is not very encouraging, although a dip in the angular distributions at small angles is reproduced. Other work in transition metals has been more concerned with the spin alignment of annihilating electrons in magnetized materials.

3.3.4. *Magnetized materials*

The polarization properties of positrons allow for investigations of the spin alignment of annihilated electrons. The interpretation of experiments with polarized positrons in ferromagnetic materials has been discussed in depth by Berko (1967).

Positrons emitted from sources such as ^{64}Cu or ^{22}Na are partially polarized parallel to their direction of motion. A substantial amount of this polarization persists throughout the thermalization until annihilation. In a magnetized ferromagnetic material two-photon annihilation of a polarized positron can only occur with electrons with opposite spin alignment. Thus only electrons in bands of appropriate spin can contribute to the photon pair momentum distribution. On magnetic field reversal the other spin bands will be sampled. Any difference between the measured momentum distributions can then be related to spin and momentum density.

Following Mijnarends (1973 b) ↓ and ↑ will denote, respectively, majority and minority electron spin directions. The relative difference

$$\frac{\Delta\Gamma(\mathbf{p})}{\Gamma_{\text{tot}}(\mathbf{p})} = \frac{\Gamma_{\downarrow}(\mathbf{p}) - \Gamma_{\uparrow}(\mathbf{p})}{\Gamma_{\downarrow}(\mathbf{p}) + \Gamma_{\uparrow}(\mathbf{p})},$$

where the $\Gamma(\mathbf{p})$ depend on the positron and spin dependent electron wavefunctions through a relation such as that of eqn. (57), will provide a measure of electron spin density at momentum $\mathbf{p}$. A full analysis of the problem (Berko 1967) exposes the following important effects :

(i) The partial positron polarization.

(ii) Three-photon decay probability.

The partial positron polarization will result in less difference between the distributions than if the positron spin alignment was complete. The effect of three-photon decay is more subtle. Any relative alignment of positron spins will cause the ratio of two-photon to three-photon events to differ from the spin-averaged value 370 (§ 1.1). The necessary renormalization of the photon pair distributions can be deduced from measurements of the probability of three-photon decay. If the experimentally measured sum and difference distributions are $R_{tot}(\mathbf{p})$ and $\Delta R(\mathbf{p})$,

$$\frac{\Delta R(\mathbf{p})}{R_{tot}(\mathbf{p})} = P_p \left\{ \frac{\Delta\Gamma(\mathbf{p})}{\Gamma_{tot}(\mathbf{p})} + P_{eff} \right\}.$$

P_p represents the average positron polarization and P_{eff} is the effective total electron polarization as seen by the positron :

$$P_{eff} = \frac{3}{4P_p} \left\{ \frac{R_+{}^{3\gamma} - R_-{}^{3\gamma}}{R_+{}^{3\gamma} + R_-{}^{3\gamma}} \right\},$$

where $R_{+(-)}{}^{3\gamma}$ is the three-photon rate for magnetic field parallel (anti-parallel) to the positron spin.

Berko (1967) discusses work prior to 1965. More recent work includes studies of nickel (Mihalisin and Parks 1966), iron (Berko and Mills 1971, Mijnarends 1973 b) and gadolinium (Hohenemser *et al.* 1968).

A qualitative interpretation of difference curves in terms of conduction electron polarization is common. However, a quantitative interpretation of the data requires computations of positron wavefunctions and electronic band structure in order that the effects of positron–electron wavefunction overlap (eqn. (57)) can be assessed. Mijnarends (1973 b) has emphasized the effect of the spin dependence of the overlap integrals in producing an *apparent* spin polarization. He has provided the most detailed comparison of positron results with band calculations.

§ 4. Defected solids and disordered systems

4.1. *The statistical approach*

4.1.1. *Positron states and lifetime spectra*

The observation of multi-component lifetime spectra can always be used to argue a variety of positron states or annihilation centres. We have already seen in chapter 2 how distinguishable positron (or positronium) states may be formed during thermalization and then may persist or be transformed during the subsequent lifetime before annihilation. In any material the number and variety of positron states may be strongly affected by the presence of defects, voids, or other centres which allow the formation of positron states that may be unlikely in more homogeneous material. Fortunately, a good understanding of how the basic characteristics of a lifetime spectrum depend on the nature and number of centres can be obtained from relatively simple models.

Most generally we may consider a system in which thermalized positrons or positronium atoms exist, in states, s, characterised by annihilation rates $\lambda(s)$, with probabilities given by a suitably normalized distribution function $P(s)$ (Brandt 1967). The form of the measured lifetime spectrum will depend on the permanence of these states relative to the mean lifetime of positrons in the system.

If each positron samples all the available states many times in an average lifetime the measured spectrum will approximate to a single component whose rate

$$\bar{\lambda} = \int \lambda(s) P(s) \, ds. \tag{73}$$

The analogy with eqn. (16) of § 1.2 is obvious.   There, we may regard the overlap of the positron wavefunction with each electron wavefunction as an annihilation state.   Single-component spectra can be expected in defected or disordered systems whenever

    (i) a weakly attractive defect potential causes only a weak local enhancement of the positron wavefunction (Connors *et al.* 1971), or

    (ii) a rapid decay of positron traps is caused by density fluctuations (Brandt 1967).

If, in the other extreme, the typical time (neglecting annihilation) for the decay of each state is very much larger than the corresponding annihilation lifetime, $\lambda(s)^{-1}$, the spectrum will take the form

$$I(t) = \int P(s) \exp\left(-\lambda(s)t\right) \, ds. \tag{74}$$

When $P(s)$ is a smoothly varying function of $\lambda(s)$ and embraces a significant range of the latter the ' annihilation rate ' will appear to decrease with time and an analysis in terms of discrete components is likely to produce confusing results.   Fortunately, in many cases, contributions to the integral will be confined to restricted ranges of $\lambda(s)$ and the spectrum will approximate to a finite number of discrete components

$$I(t) = \sum_{i}^{N} I_i \exp\left(-\lambda_i t\right). \tag{75}$$

Then the component intensities are proportional to the initial state populations and the rates can be associated with the effective electron density in each state.   Some such components will always arise when essentially stationary positron states are formed during the thermalization process.

More realistic are those cases in which some transitions between the various states can occur and be described, with an obvious notation, by time independent rates $K_{ij}(K_{ij} \neq K_{ji}; \quad K_{ii}=0)$.   Any initial positron population then decays as

$$n(t) = \sum_{i}^{N} n_i(t), \tag{76}$$

where the $n_i(t)$ satisfy a set of coupled differential equations :

$$\frac{dn_i(t)}{dt} + \left(\lambda_i + \sum_{j}^{N} K_{ij}\right) n_i(t) = \sum_{j}^{N} K_{ji} n_j(t). \tag{77}$$

The lifetime spectrum will take the form

$$I(t) = \sum I_\nu \exp\left(-\Gamma_\nu t\right), \tag{78}$$

where the I_ν and Γ_ν in general will be complicated functions of the λ_i and K_{ij}. Analytic solutions of eqn. (77) are possible when N is small but even then, they are complicated, and approximations are useful.

If $K_{ij} \gg \lambda_i$ for all i, j, approximate solutions analogous to eqn. (73) can be obtained (Bertolaccini and Dupasquier 1970, Connors *et al.* 1971) and, for example, in the two-state case

$$n(t) \simeq n_0 \exp \left[-\left(\frac{\lambda_1 K_{21} + \lambda_2 K_{12}}{K_{21} + K_{12}} \right) t \right]. \tag{79}$$

The discussion following eqn. (73) again applies.

The simple 'trapping' models (Brandt 1967, Bergersen and Stott 1969, Connors and West 1969) assume that, because of the rapidity of thermalization in most materials, at $t = 0$ all the positrons lie in a single state from which they either annihilate or make transitions to other states or regions of different electron density. If the probability of subsequent transitions is vanishingly small,

$$\frac{dn_1(t)}{dt} + \left(\lambda_1 + \sum_j K_{ij} \right) n_1(t) = 0,$$

$$\frac{dn_j(t)}{dt} + \lambda_j n_j(t) = K_{1j} n_1(t)$$

for all $j \neq 1$. The decoupling of the equation for $n_1(t)$ allows an easy solution which with the boundary conditions $n_j(0) = n_0 \delta_{1j}$, provides a spectrum

$$n(t) = n_0 \sum_j I_j \exp \left(-\Gamma_j t \right),$$

where

$$n(t) = n_0 \left\{ 1 - \sum_j \frac{K_{1j}}{(\lambda_1 - \lambda_j + \Sigma)} \right\} \exp \left[-(\lambda_1 + \Sigma)t \right]$$

$$+ \sum_j \frac{n_0 K_{1j}}{(\lambda_1 - \lambda_j + \Sigma)} \exp \left(-\lambda_j t \right). \tag{80}$$

Here, $\Sigma \equiv \sum_j K_{1j}$.

Thus, in this model, the decay rate of any resolvable component, $j \neq 1$, is equal to the annihilation rate in the corresponding state or centre. For the first state

$$\lambda_1 = \sum_j I_j \Gamma_j \tag{81}$$

(Bertolaccini *et al.* 1971).

The simple two-state trapping model has proved particularly useful in a large variety of problems.   Here

$$n(t) = n_0 \left\{ 1 - \frac{K_{12}}{(\lambda_1 - \lambda_2 + K_{12})} \right\} \exp\left[-(\lambda_1 + K_{12})t \right]$$

$$+ \frac{n_0 K_{12}}{(\lambda_1 - \lambda_2 + K_{12})} \exp\left(-\lambda_2 t \right). \qquad (82)$$

More complex patterns of transitions can be envisaged.   Bertolaccini *et al.* (1971) have considered a three-state model with K_{12}, K_{23}, K_{13} non-zero.   Detailed analyses of the mechanisms of positron trapping and their effect on the lifetime spectra have been applied to two and three-state models by Brandt and Paulin (1972).   Still more complex solutions no doubt pertain to many systems.   Even so, it is extremely unlikely that the present uncertainties in experiment and analysis will allow their detection.

4.1.2. *Analysis of experiments*

The analysis of experimental lifetime results for multi-component spectra is difficult.   The convolution of a real spectrum with an impre-cisely defined time resolution curve of finite width causes significant distortion.   Even with infinitely good resolution, a reliable analysis of data of limited statistical accuracy in terms of sums of exponentials is far from easy (Lanzcos 1957).   Superficially, the problem would appear to be eased by recourse to theoretical models for both the time resolution curve and the real spectrum (Lichtenberger *et al.* 1971, Snead *et al.* 1972). Application of the usual significance tests is essential.   Even then, one can never be sure to what extent the derived results are model-dependent.

Practicality demands that we set our forebodings aside.   We therefore turn to the problem of the interpretation of derived component intensities and rates in situations where the simpler trapping models appear to be applicable.

Observable effects in lifetime spectra can occur whenever a K_{1j} is of the same order as λ_1 (eqn. (80)).   If this condition obtains at a sufficiently low concentration of traps, c_j, an equation of the form

$$K_{ij} = \nu_{1j} c_j \qquad (83)$$

should be applicable.   A change in the concentration of traps can then be followed through the consequent changes in Γ_1, or the appropriate component intensity.

The interpretation of Γ_1 as $\lambda_1 + \Sigma$ (eqn. (80)) is only valid for times sig-nificantly greater than the thermalization time, although a more practical limitation will generally result from the disappearance of this component into the time resolution curve.   Measurements of the variations in com-ponent intensity are also limited by practical considerations but the func-tional dependence of I_j on c_j, when deducible, provides an important guide to the processes by which that positron state obtains (Dupasquier 1970).

An analysis of the lifetime spectrum always provides the maximum amount of quantitative information about positron trapping phenomena. Nevertheless, the measurement of other parameters can be useful. The mean lifetime, angular correlation, or annihilation line shape will be determined by the total number of annihilations in each state :

$$N_j = \int_0^\infty \lambda_j n_j(t) \, dt.$$

For the trapping model (eqn. (80))

$$N_1 = \frac{\lambda_1 n_0}{(\lambda_1 + \Sigma)}, \quad N_j = \frac{K_{1j} n_0}{(\lambda_1 + \Sigma)}. \tag{84}$$

In general $I_j \leqslant N_j$ and only if $\lambda_j \ll \lambda_1$, as in the case of trapping in large voids will the equality hold.

If any characteristic of annihilation, F, is a linear function of the positron state in the sense that

$$\bar{F} = \sum_j F_j N_j / \sum_j N_j, \tag{85}$$

then its measurement can, in principle, provide quantitative information about state populations or trap concentrations. It is essential that F be a good characteristic in that (a) it exists for all states j, (b) it does not represent the more subtle features of the annihilation process, and (c) it can be measured with comparative precision. Measurable parameters that may be good and linear in the sense described above are

(i) the mean lifetime t,
(ii) the ratio of peak height to area of a long slit angular distribution, h (Connors *et al.* 1971),
(iii) the peak height to area ratio of a doppler broadened energy curve, S (Mackenzie *et al.* 1970).

One example of a defect sensitive but non-linear characteristic is the width of an angular distribution.

The linear properties (eqn. (85)) of positron lifetime are an essential assumption of the models of § 4.1.1. Such an assumption is likely to be more fragile in the case of h (or S) which is more sensitive to the detailed form of the electron and positron states. Nevertheless, the investigation of the relationship between defect-induced changes in t and h or S will always provide a useful test of the basic philosophy of independent and well-defined positron states.

Although in many materials an adequate measure of t, h, or S may be obtained more easily and quickly than a lifetime spectrum of sufficient precision for multi-component analysis, the information obtained is necessarily more limited. Thus similar changes in t can result from radically different types of spectra (Connors *et al.* 1971, McKee, Triftshäuser and Stewart 1972). However, changes in angular correlation can sometimes provide additional information not deducible from lifetime spectra.

The various ways in which the changes in angular distribution can be related to the particle states and nature of positron traps will be described in following sections.

4.2. *Ionic crystals*

4.2.1. *Positron states in ionic materials*

We have already noted in § 2.3 the observation of complex lifetime spectra for alkali halides (Bisi *et al.* 1964), the studies of magnetic quenching (Bisi *et al.* 1966, Gainotti *et al.* 1963) and the hypotheses of various types of positron state or annihilation centre (Goldanskii and Prokop'ev 1965, 1966, Prokop'ev 1966, Brandt 1967). Bussolati *et al.* (1967) re-measured the lifetime spectra of alklai halides using improved time resolution. All their spectra could be resolved into at least two components which were tentatively assigned to the annihilation of the ground and first excited states of the positron–negative ion bound system proposed by Goldanskii *et al.* A modification to this interpretation was suggested by Bertolaccini and Dupasquier (1970) who investigated both solid and liquid alkali halides. Abrupt changes in the form of the lifetime spectra at melting were used as evidence against the bound state hypothesis and positron trapping by defects was argued for the origin of the longer-lived components. Trapping model analyses were used to explain a decrease in component number on melting, an effect which is in accord with the qualitative analysis given in § 4.1.1.

More recent studies by Bertolaccini *et al.* (1971) have revealed the presence of at least two significant components in the spectra for a larger number of analytical reagent grade polycrystalline and single-crystal ionic solids. Application of the two-state trapping model (eqn. (82)) to those spectra containing two significant components disclosed a correlation between annihilation rates and crystal properties which was not apparent from an analysis based on eqn. (75). A connection between the annihilation rate λ_1 (table 4) and crystal properties was established with the aid of the following model.

The volume of the crystal occupied by positive ions is considered to be inaccessible to positrons. In the remaining accessible regions of the crystal the ' density ' of negative ions is taken to be

$$n^* = \frac{n v_-}{1 - \frac{4}{3}\pi R_+{}^3 n v_+}.$$

Here, $n = \rho N_0/M$, ρ is the density, N_0 is Avogardo's number, M is the molecular weight, R_+ is the radius of the positive ion, and v_- and v_+ are, respectively, the number of negative and positive ions per molecule. A plot (fig. 32) of λ_1 versus n^* discloses a linear dependence of the form

$$\lambda_1 = \lambda_0 + k n^*.$$

The deduced value for λ_0 ($1 \cdot 95 \pm 0 \cdot 8$ nsec^{-1}) is close to the spin-averaged

Table 4. Annihilation rates λ_1, λ_2, and transition rates K_{12}, for the first two positron states in various ionic media. n^* as in fig. 32 (Bertolaccini *et al.* 1971).

	n^* (10^{22} cm^{-3})	λ_1 (10^{10} sec^{-1})	λ_2 (10^{10} sec^{-1})	k_{12} (10^{10} sec^{-1})
Cu_2O	3·11	0·341 ± 0·031	0·171 ± 0·005	0·017 ± 0·008
ZnO	4·46	0·359 ± 0·019	0·259 ± 0·008	0·317 ± 0·038
PbO_2	5·01	0·434 ± 0·027	0·317 ± 0·010	0·137 ± 0·024
ZrO_2	5·71	0·454 ± 0·038	0·222 ± 0·007	0·078 ± 0·010
SnO_2	5·80	0·451 ± 0·038	0·216 ± 0·006	0·062 ± 0·008
NiO	6·63	0·532 ± 0·044	0·283 ± 0·008	0·074 ± 0·010
MgO	5·91	0·481 ± 0·045	0·189 ± 0·006	0·012 ± 0·002
CaO	4·13	0·386 ± 0·035	0·149 ± 0·005	0·017 ± 0·002
SrO	3·41	0·360 ± 0·032	0·165 ± 0·005	0·035 ± 0·005
CaS	2·30	0·359 ± 0·011	—	—
SrS	2·15	0·368 ± 0·011	—	—
BaS	1·83	0·318 ± 0·010	—	—
CaSe	1·96	0·373 ± 0·034	0·172 ± 0·005	0·017 ± 0·002
SrSe	1·79	0·352 ± 0·011	—	—
BaSe	1·67	0·324 ± 0·011	—	—
SrTe	1·50	0·326 ± 0·011	—	—
BaTe	1·35	0·295 ± 0·028	0·11 ± 0·003	0·004 ± 0·003
LiF	7·77	0·482 ± 0·024	0·337 ± 0·010	0·276 ± 0·037
NaF	4·87	0·384 ± 0·018	0·324 ± 0·009	0·134 ± 0·028
KF	3·59	0·345 ± 0·021	—	—
RbF	3·27	0·317 ± 0·018	—	—
CsF	2·91	0·351 ± 0·025	—	—
LiCl	3·27	0·329 ± 0·021	0·230 ± 0·007	0·095 ± 0·016
NaCl	2·63	0·269 ± 0·020	0·146 ± 0·004	0·050 ± 0·006
KCl	2·06	0·275 ± 0·019	0·159 ± 0·005	0·089 ± 0·012
RbCl	1·87	0·296 ± 0·019	—	—
CsCl	2·32	0·327 ± 0·024	—	—
LiBr	2·62	0·310 ± 0·025	0·180 ± 0·006	0·027 ± 0·004
NaBr	2·15	0·309 ± 0·025	0·134 ± 0·004	0·033 ± 0·004
KBr	1·73	0·250 ± 0·020	0·132 ± 0·004	0·038 ± 0·005
CsBr	1·90	0·382 ± 0·033	—	—
LiI	1·66	0·286 ± 0·206	0·152 ± 0·005	0·020 ± 0·002
NaI	1·64	0·287 ± 0·026	0·129 ± 0·004	0·018 ± 0·002
KI	1·35	0·285 ± 0·020	—	—
RbI	1·23	0·275 ± 0·017	—	—
CsI	1·46	0·341 ± 0·025	—	—
MgF_2	6·23	0·490 ± 0·036	0·327 ± 0·010	0·098 ± 0·018
CaF_2	5·51	0·447 ± 0·033	0·305 ± 0·009	0·112 ± 0·019
SrF_2	4·77	0·429 ± 0·033	0·288 ± 0·008	0·146 ± 0·023
BaF_2	4·16	0·380 ± 0·027	0·265 ± 0·008	0·188 ± 0·027
$BaCl_2$	2·58	0·313 ± 0·028	0·175 ± 0·005	0·028 ± 0·005
$CaBr_2$	2·12	0·242 ± 0·018	—	—
$SrBr_2$	2·22	0·290 ± 0·022	0·186 ± 0·006	0·078 ± 0·012
SrI_2	1·70	0·284 ± 0·024	—	—

positronium rate or the charactristic rate for a negative positronium ion *in vacuo* ($\lambda_{Ps^-} = 1\cdot99$ nsec^{-1}) (Ferrante 1968). The value for k is appropriate to the spin-averaged rate for $\sim$six electrons per ion. Following these authors, we may note that the results are indicative of a positronium-like atom or negative ion which is closely associated with the negative ions of the containing system.

Fig. 32

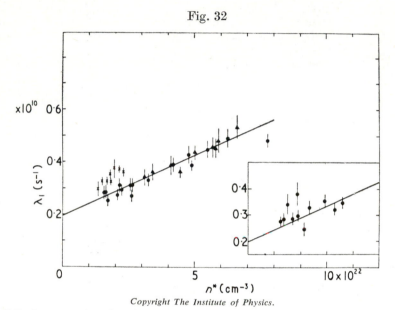

Annihilation rate for the shortest lived positron state in various ionic solids. The experimental data is taken from table 4. n^* is the 'density' of negative ions as seen by the positron (Bertolacini *et al.* 1971).

A similar study by the same authors (Bertolaccini *et al.* 1971) of a possible correlation between crystal properties and the second-component rate λ_2 (table 4) is less conclusive. λ_2 appears to depend on the properties of both the positive and negative ions. The applicability of the two-state trapping model suggests that this state is filled from the first positron state and positron trapping by defects is plausible.

Bisi, Dupasquier and Zappa (1971 a) have given a brief report of ortho-positronium-like magnetic quenching in nominally pure NaCl and LiF crystals. Further evidence of positronium states in ionic solids is provided by the observation of the narrow peak, symptomatic of paraposi-tronium self-decay, in the angular distributions for single crystals of quartz (Coussot 1969). Subsequent studies in quartz (Brandt *et al.* 1969, Greenberger *et al.* 1970, Hodges *et al.* 1972), CaF$_2$ (Brandt *et al.* 1969), and ice (Mogensen *et al.* 1971) are made more striking by the simultaneous observation of narrow satellite peaks which appear at the projections of reciprocal lattice vectors onto the measured momentum direction (fig. 33).

The common parapositronium, origin of the satellite and central peaks in quartz, has been confirmed by magnetic quenching studies (Greenberger *et al.* 1970) which show a similar enhancement of the intensity of all the peaks. Irradiation with electrons (Coussot 1969) or by repeated positron studies (Hodges *et al.* 1972) decreases the narrow component intensities which also depend strongly on the crystal origin (fig. 33). Lifetime studies by Bisi, Gambarini and Zappa (1971) have shown that the enhanced intensity of narrow components in the angular distributions for synthetic

Fig. 33

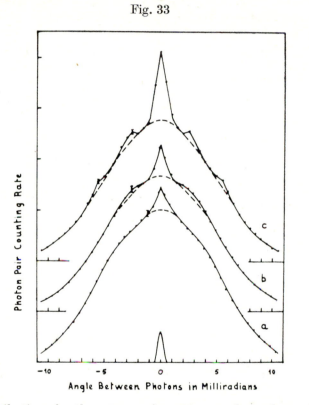

Angular distributions for three types of quartz crystals as observed along the
Y-crystal direction. (*a*) Natural quartz from Brazil. (*b*) Natural
quartz from Madagascar. (*c*) Synthetic quartz. (Brandt *et al.* 1969)

quartz crystals is paralleled by the increased ($\sim 40\%$) intensity of a second longer-lived lifetime component who positronium-like origin is confirmed by magnetic field quenching. Application of the trapping model result (eqn. (81)), provides a similar value for the annihilation rate, λ_1, of the shorter-lived positron state in three different samples of natural and synthetic quartz, notwithstanding radically different I_2. Bisi *et al.* (1971) conclude that the observed effects in angular correlation and lifetime arise from positron transitions, from an initial thermalized state common to all crystals, to a positronium state trapped in crystal defects.

An explanation of the satellite parapositronium peaks in angular distributions has been given by Hodges *et al.* (1972). The satellite peaks are shown to be a necessary consequence of HMC in the parapositronium decay which arises from the orthogonalization of the positronium and valence electron wavefunctions. A theoretical calculation gives good agreement with the peak intensities in their own angular distributions for quartz (Hodges *et al.* 1972) and the somewhat smaller satellite peaks in ice (Mogensen *et al.* 1971). The small width of the parapositronium peaks suggests delocalized positronium states (Hodges, private communication). At first sight this result would seem to conflict with the defect trapping picture presented by Bisi, Gambarini and Zappa (1971). However it may be that the defects merely serve as centres for the formation of positronium atoms which subsequently propagate freely through the crystal.

4.2.2. *Positron traps in alkali halides*

The strongest evidence of positron trapping in alkali halides has come from studies in which established methods (Schulman and Compton 1962) have been used to induce additional defects or positron traps.

Williams and Ache (1969) found an increased intensity of long-lived component in the spectra for proton and gamm-irradiated NaCl and NaF single crystals. Thermal annealing reduced the intensity of the defect-induced components. The severity and variability of the irradiation process and the limited resolution of the lifetime measurements did not allow for the identification of the positron traps.

A more restrained treatment of samples by Brandt *et al.* (1968, 1971) established a strong correlation between spectral intensities and V-centre absorption (Schulman and Compton 1962, Pick 1972). Lifetime spectra in NaCl crystals were resolved into two components. The optical absorption and lifetime spectra were measured after various sequential stages of sample preparation. An initial concentration of vacancies, produced by cleaving the specimens, was assessed to be ~ 1 p.p.m. and gave $I_2 \sim 50\%$. Low dose room temperature irradiation with $50\,\mathrm{kV}$ X-rays caused progressive conversion to F and V-centres and decrease in I_2. A further stage of optical bleaching preferentially destroyed the F-band absorption without affecting the lifetime spectrum. A final stage of thermal annealing at $480°\mathrm{C}$ removed most of the V-region colouring and increased I_2 to its initial value.

Changes in I_2 were related to changes in concentration of the effective positron traps by the relation

$$\Delta_2 = \delta \left(\frac{I_2}{1 - I_2} \right) = \frac{\nu_{12} \delta c_2}{\lambda_1 - \lambda_2} \tag{86}$$

which is easily deduced from eqn. (82). A linear relation between $-\Delta_2$ and colour-centre concentration (fig. 34) suggests positron trapping by potential V-centre defects. At temperatures $> 220°K$ the most stable V-centre is believed to be the V_F-centre or positive-ion vacancy + trapped hole (Pick 1972). These considerations led the authors (Brandt *et al.* 1971) to interpret their results in terms of a trapping ability of positive-ion vacancies which can be rendered inoperative by prior hole capture.

Fig. 34

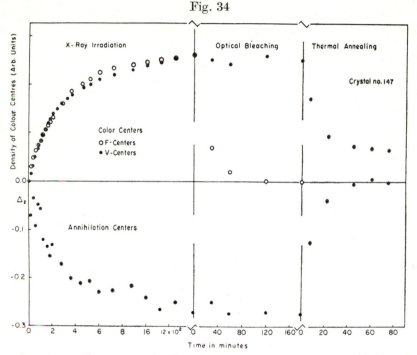

Growth and annealing curves for F and V-centre concentrations in NaCl and the change in I_2 as expressed by Δ_2 (eqn. (86)) (Brandt *et al.* 1971).

At sufficiently high concentrations the effect of positron trapping by F-centres is also observable. Dupasquier (1970) measured positron lifetimes in additively and electrolytically coloured KCl crystals having F-centre concentrations up to $\sim 10^{18}$ cm^{-3}. A three-component analysis of the lifetime spectra obtained after various types of crystal pretreatment revealed the following pattern. The slowest component had an intensity, I_3, which grew with increasing F-centre concentration, and a decay rate ~ 1.05 nsec^{-1} independent of specimen history. Increases in I_3 were accompanied by increases in the rates, Γ_1, Γ_2, of the remaining components. Pretreatment to produce excess potassium colloidal centres was found to have little effect on the spectra. A trapping model analysis (Bertolaccini *et al.* 1971) suggests that the longest-lived state is filled both by direct transitions from the shortest-lived state and indirect transitions via the state represented by the intermediate component.

Dupasquier attributed the slowest component to the pick-off decay of orthopositronium atoms bound in the anion vacancies. Angular correlation studies (Herlach and Heinrich 1970 a, b, Arefiev and Vorobiev 1972) support this hypothesis. The angular distributions for additively coloured KCl crystals are narrower than those for 'pure' KCl. This narrowing can be understood in terms of the annihilation of trapped positrons with F-centre electrons having comparatively small momenta (Herlach and Heinrich 1970 a). The detection of positronium-like states in additively coloured KCl by magnetic quenching studies (Bisi, Dupasquier and Zappa 1971 b) may not be relevant in view of similar results in nominally pure systems (Bisi, Dupasquier and Zappa 1971 a).

A comprehensive interpretation of these various results is made difficult by the often but inevitable lack of precise definition of the state of the sample, variations in the number of components resolved, and differing choices of trapping model. However, a significant step in this direction has been made by Mallard and Hsu (1972) who also consider their own investigation of lifetime spectra for KCl (fig. 35).

Fig. 35

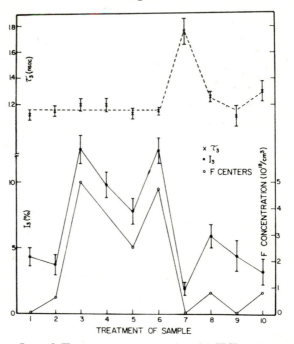

Variation of τ_3, I_3 and F-centre concentration in KCl : (1) as cleaved ; (2) $2 \cdot 2 \times 10^6 \mathrm{R}$; (3) $588 \times 10^6 \mathrm{R}$; (4) optically bleached 2 hours ; (5) optically bleached 24 hours ; (6) $2 \cdot 2 \times 10^6 \mathrm{R}$; (7) annealed at $250°\mathrm{C}$; (8) $2 \cdot 2 \times 10^1 \mathrm{R}$ (9) annealed at $450°\mathrm{C}$, (10) $2 \cdot 2 \times 10^6 \mathrm{R}$ (Mallard and Hsu 1972).

These authors suggest three competing processes for the origin of the longest-lived component.

A. Annihilation associated with free positive ion vacancies.

B. Annihilation associated with F-centres.

C. Annihilation associated with other defects generated by radiolysis, deformation, or heat treatment.

An abbreviated account of their analysis of their results (fig. 35) now follows.

The initial value of I_3 is mainly determined by process A, (1). Low dose radiation produces isolated F-centres and halide interstitials which are anchored by positive ion vacancies with consequent decrease in the free positive ion vacancy concentration. The decrease in I_3, at (2), where the F-centre and vacancy concentrations are similar ($\leqslant 5 \times 10^{17}$ cm^{-3}), is due to a greater probability of trapping by positive ion vacancies than by F-centres (at the same concentration). The data of fig. 34 support this assertion.

At higher F-centre concentrations (3), (4), (5), (6) process B dominates and I_3 correlates well with F-centre concentration as in the measurements of Dupasquier (1970).

The 250°C anneal (7) ionizes the F-centres without breaking up the halogen clusters. Processes A and B are both suppressed and a residual component which, if not source, is attributable to other defects, becomes apparent. At (8), τ_3 and I_3 are larger than at (2) as a result of incomplete defect annealing. Further annealing at 450°C returns the crystal to its initial condition.

For all three components the values of lifetime obtained by Mallard and Hsu are roughly in agreement with the measurements of Dupasquier, although in contrast with those measurements, τ_1 and τ_2 appear constant throughout the cycle of treatment. However, this, and the apparent linear dependence of I_3, on F-centre concentration, do not rule out the trapping model. Inadequacies in spectrum analysis could easily explain the non-observation of the comparatively subtle differences between the various theoretical models when I_3 is small.

The trapping effect of F-centres would appear to be clear. Further and convincing evidence of trapping by positive ion vacancies has been supplied by Brandt and Waung (1971) with lifetime studies of Ca^{2+} doped crystals of KCl.

Divalent impurity doping produces a concentration of free positive-ion vacancies, n_v, proportional to the square root of the impurity concentration C. A quantitative relation (Lidiard 1957),

$$n_v \simeq (C/Z_1)^{1/2} \exp\left[-E_a/k_B T\right] \tag{87}$$

involves the positive ion coordination number Z_1, and the free energy of association, E_a between an impurity ion and a positive ion vacancy on adjacent sites. By an analysis based on a three-state trapping model (K_{12}, K_{13} non-zero-eqn. (80)) Brandt and Waung deduced that $K_{13} \propto C^{1/2}$ consistent with trapping by positive-ion vacancies.

Other more qualitative studies of the effects of impurity doping in various ionic solids reveal variations in lifetime (Singh *et al.* 1970, Hsu *et al.* 1971, Tumosa *et al.* 1971, Kelly 1971), and Doppler broadening (Kelly 1971) that are generally consistent with contemporary beliefs about the effects of the dopants on potential positron traps.

A comparison of the more quantitative results of various experiments is complicated by the effects of the practical problems of spectrum analysis discussed in § 4.1.2. Essentially similar data (Brandt and Waung 1971, Dupasquier 1970) may be studied with essentially different models (Brandt and Waung 1971, Bertolaccini *et al.* 1971).

Our conscience eased by this re-iteration of earlier expressed fears (§ 4.1.2) we turn our attention to the longest-lived lifetime component for KCl. A positronium-like nature for the state that results from F-centre capture is most plausible and a theoretical estimate of the lifetime of this state should not be too difficult a task for the theorist. The similar lifetime of the state associated with positive-ion vacancies in KCl is probably irrelevant. However, the effect of ionic volumes should be an important factor in determining the lifetime in both types of trap and comparative studies in different ionic solids would be of value.

In some cases the resolution of the individual lifetime components may not be possible. Such may be the case in NaCl (table 5) where the apparent increase in Γ_2 with I_2 suggests inadequate resolution of at least two components. Tumosa *et al.* (1971) have encountered difficulties in applying the simple two-state model to spectra for NaCl.

Application of the trapping model relation (eqn. (81)) to the data of table 5 gives reasonably consistent values for the annihilation rate of the shortest-lived state in both ' pure ' and defected crystals. The observed values of Γ_2 in KCl are again not inconsistent with that for the pure crystal ($\Gamma_2 = \lambda_2$) if the three-state model of Bertolaccini *et al.* (1971) is applicable. However the origin or origins of this intermediate component remains unclear.

For a final word on the identification of the positron states in KCl we turn to the interesting numerical analysis made by Brandt and Waung (1971). These workers assume that the thermalized positrons interact strongly with negative ions and diffuse by random walk from anion to anion until they annihilate or are captured by positive-ion vacancies at a rate

$$K_{1v} = 4\pi R D_+ \Omega^{-1} n_v.$$

Here Ω is the volume of a unit cell, n_v is the concentration of positive-ion vacancies, R is the effective capture radius ($\sim a/2$ where a is the lattice constant), and D_+ is the positron diffusion coefficient. A numerical analysis of results for Ca^{2+}-doped KCl crystals (see eqn. (87) and the associated discussion) leads to an estimate for the diffusion coefficient $D_+ \approx 6 \times 10^{-3}$ cm^2 sec^{-1}. Such a value implies a dwell time in each unit cell, $\tau \approx 3 \times 10^{-14}$ sec. This is some 100 times longer than the eigentime

Table 5. Lifetime spectrum intensities and rate for NaCl and KCl crystals. Data marked * have been deduced from that given in the original references. The values in the penultimate column were obtained from eqn. (81).

Material and reference	Γ_1 (10^9 sec^{-1})	I_1	Γ_2 (10^9 sec^{-1})	I_2	Γ_3 (10^9 sec^{-1})	I_3	$\lambda_1 = \sum_\nu \Gamma_\nu I_\nu$	Supposed positron traps
NaCl								
Bertolaccini et al. (1971)	3·2*	0·71*	1·5	0·29*	—	—	2·7	'Pure' crystals
Williams and Ache (1969)	2·9	0·81	1·2	0·19	—	—	2·5	?
Brandt et al. (1971)	4·6*	0·50	2·2	0·50	—	—	3·4	Cation vacancies
Tumosa et al. (1971)	2·5	0·96	1·0	0·04	—	—	2·5	'Pure' crystals
Tumosa et al. (1971)	3·6	0·60	1·8	0·40	—	—	2·8	Cation vacancies
KCl								
Bertolaccini et al. (1971)	3·6*	0·57*	1·6	0·43*	—	—	2·8	'Pure' Crystals
Mallard and Hsu (1972)	4·8	0·27	2·0	0·61	0·9	0·12	3·3	F-centres
Dupasquier (1970)	8·3	0·23	2·3	0·54	1·1	0·21	3·3	F-centres
Brandt and Waung (1971)	4·4	0·35	2·1	0·46	0·9	0·19	2·7	Cation vacancies

$\hbar/E = 2 \times 10^{-16}$ sec of a Wheeler compound, Cl^-e^+ of binding energy, $E = 3 \cdot 74$ eV, and thus sufficiently long for the positron to form a quasi-stable bound state with the negative ion.

An analysis of the second component rate K_{12}, using the deduced value for D_+ implies a value for the product of trap concentration and capture radius considerably greater than in the vacancy case. This, and the intermediate value of the corresponding annihilation lifetime ($\lambda_2^{-1} \approx 0 \cdot 6$ nsec), suggests positron traps which bind a positron rather less tightly than a negative ion ($\lambda_1^{-1} \approx 0 \cdot 3$ nsec) or a positive-ion vacancy ($\lambda_3^{-1} \simeq 1 \cdot 2$ nsec) (table 5). Such an interpretation of the intermediate state would certainly allow for subsequent transitions to longer-lived F-centre states as suggested by Bertolaccini et al. (1971).

At present, positron studies can do little more than lend support to the more established methods of colour-centre research. Increased precision in lifetime studies (table 5) would be an important step towards a realization of a somewhat greater potential.

4.3. Metals

4.3.1. Experiments

The effect of deformation on the angular distribution for a metallic system was first reported by Dekhtyar et al. (1964). A narrowing of the angular distributions of Fe–Ni alloys was attributed to changes in electronic structure brought about by the deformation. At about the same time, MacKenzie et al. (1964) reported the similar and pronounced effects

Table 6. Angular distribution changes on heating and melting. Data from Kusmiss (1965)

Metal	$\dfrac{\text{core area}}{\text{total area}} = U$	ΔU-heating	ΔU-melting
Hg	0·71	0·00	−0·11
Zn	0·48	−0·16	−0·01
In	0·43	−0·11	+0·01
Ca	0·39	−0·08	+0·01
Pb	0·40	−0·06	0·00
Tl	0·38	−0·05	+0·02
Sn	0·36	−0·04	—
Bi	0·31	+0·04	−0·05
Li	0·21	−0·03	0·00
Sb	0·20	−0·01	−0·02
Te	0·16	−0·02	−0·02
Al	0·19	−0·10	+0·01

of temperature on the angular distributions of zinc, cadmium and indium. Kusmiss and Stewart (1967) investigated the effect of temperature and melting on the angular distributions for 22 metals (table 6). The narrowing of the angular distribution, which in some metals occurs on heating the solid and in others on melting, was identified as a relative increase in the proportion of the parabolic part of the distribution accompanied by a smearing of the discontinuity between parabola and gaussian (fig. 36).

Fig. 36

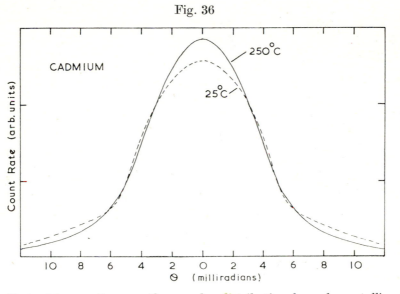

The effect of temperature on the angular distribution for polycrystalline Cd.

Noting these results, Berko and Erskine (1967) investigated the angular distributions for plastically deformed samples of aluminium and suggested that the magnitude of the observed changes in angular distribution shape could be explained by a positron affinity for regions, such as dislocations or vacancies, of lower than average electron density. The necessary modification in positron lifetime was observed by Grosskreutz and Millet (1969) in studies of deformed aluminium and copper. The observed changes in lifetime were considerable and again suggested a marked positron sensitivity to defects.

A pointer to the origin of some of these changes was provided by the lifetime studies of MacKenzie *et al.* (1967) who measured the temperature dependence of positron lifetimes in several solid metals. In Zn, Cd, and in (fig. 37) the results were sufficient to show a smooth change in mean lifetime between essentially constant, initial low temperature and final high temperature values. Lack of hysteresis in the measurements suggested equilibrium vacancy concentration as responsible for the effect.

A realization of the potential of such studies in the measurement of vacancy formation energies resulted from the demonstration of the applicability of a simple two-state trapping model (Brandt 1969, unpublished conference, Bergersen and Stott 1969, Connors and West 1969). Section 4.3.3 is our response to the enthusiasm (Seeger 1973) which has greeted this development.

Fig. 37

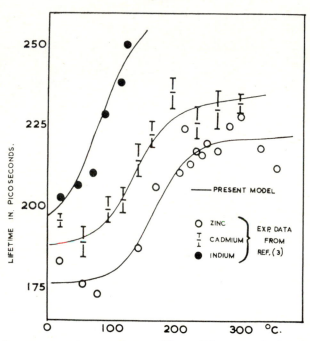

Positron lifetimes versus temperature. The solid curves derive from the two-state trapping results, eqns. (91), (92). Here Ref. (3) refers to Mackenzie *et al.* (1967) (Connors and West 1969).

An early but crude attempt at quenching experiments (Connors *et al.* 1970) involved cadmium samples 'quenched' from a temperature of 570°K into a liquid nitrogen bath. The angular distribution of the cold (93°K) sample had the form characteristic of a sample temperature of 570°K. Subsequent experiments involving improved quenching techniques (fig. 38) clearly demonstrate irreversible annealing effects. The results of more controlled treatment of aluminium samples have been reported by Cotterill, Petersen, Trumpy and Träff (1972). Quenching and annealing procedures allowed some separation of the effects of vacancies and of dislocation loops. Positron lifetimes and trapping rates were deduced for each type of defect.

Deformation studies are more numerous. The effect of deformation on positron lifetime in aluminium (Grosskreutz and Millet 1969, Hautojärvi *et al.* 1970, Cotterill, Petersen, Trumpy and Träff 1972) is some 40% increase in lifetime. The corresponding changes in angular distribution parameters (see § 4.1.2) are smaller (Berko and Erskine 1967, Hautojärvi and Jauho 1971).

Fig. 38

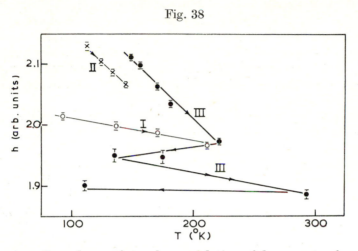

Temperature dependence of angular correlation of *h*-parameter for cadmium samples initially quenched from 570°K to liquid nitrogen. I (Connors *et al.* 1970), II and III (Connors unpublished), refer to different initial quenches. The irreversibility of post-quench temperature-induced changes can be seen in III.

In deformed copper the changes in angular correlation (Hautojärvi and Jauho 1971, Kuribayashi *et al.* 1972) are of a similar magnitude to those in positron lifetime (Grosskreutz and Millet 1969). A comparison with the corresponding changes in Doppler-broadened lineshape (MacKenzie *et al.* (1970)) is complicated by resolution effects. Annealing studies (Kuribayashi *et al.* 1972) indicate that most of the deformation-induced change in angular correlation is caused by clustered vacancies and dislocations.

Kusmiss *et al.* (1970) have also attempted to distinguish between the effects of vacancies and other defects. A narrowing of the angular distributions for platinum samples, arising from vacancy-inducing prior electron irradiation, was compared with the similar but more pronounced narrowing caused by deformation. Snead *et al.* (1971) have reported similar relative effects in iron. A happy association of positron and defect experts provides an extensive analysis and interpretation which is guided by the results of more conventional defect studies. Particularly interesting are the analyses of defect-induced changes in angular distribution shape and the observation of the inhibition of single-vacancy trapping by carbon atom impurities.

The work of Dekhtyar and his colleagues should be mentioned. Dekhtyar (1970) has considered the two-state model for positron trapping by defects, but his equation for the fraction of positrons trapped radically differs from our own (eqn. (84)). Consequently we find his conclusions (Dekhtyar 1970, 1971, Dekhtyar and Cisek 1971) hard to assess. Several references to work by Dekhtyar and his colleagues are given by Adamenko and Shalaev (1971) in an interesting report of angular correlation studies in gamma-irradiated tungsten crystals. The effect of the irradiation was to broaden the angular distribution, a result attributed to annihilation in regions of the metal where the ionic density is increased through the presence of interstitials. Reference is made to similar observations in quenched metal.

The curves of fig. 36 can be taken as representative of defect-induced changes in angular distributions. The magnitude of these changes, and the usually larger changes in lifetime, must cast some doubt on positron results obtained prior to the recognition of positron trapping effects. MacKenzie, Craig and McKee (1971) have noted that the positron vacancy-trapping phenomenon could be responsible for part or all of the temperature-induced smearing effects in alkali metal angular distributions. However, the credibility of the effective mass interpretation of this smearing (§ 2.4.4) is sustained by their investigation of lifetime changes which show less than 3% variation from liquid helium to melting. These results and theoretical considerations (Seeger 1972) suggest that positron defect-trapping phenomena is not very likely in alkali metals.

4.3.2. ' Theories '

A general discussion of the experiments and phenomena described in § 4.3.1 can be conveniently divided into three important, but not mutually exclusive, parts.

    (i) The applicability of the various models of § 4.1.2.
    (ii) The mechanisms of positron trapping by defects.
    (iii) The nature of the defect-induced positron states.

(i) *Models*

An assessment of the relative merits of particular solutions of eqn. (77) such as enhancement (eqn. (79)) or trapping models (eqn. (80)) can only be made from an analysis of the lifetime spectra. However, the resolution of spectral components of similar rates and of the rapidity characteristic of metals represents the most demanding test of present techniques.

Unconstrained four parameter (I_1, Γ_1, I_2, Γ_2) analyses of lifetime spectra for hot (McKee, Jost and MacKenzie 1972) and deformed aluminium (Hautojärvi *et al.* 1970) give some evidence of the superiority of a two-component fit. Striking demonstrations of the applicability of the trapping models to vacancy effects in aluminium are provided by the deduction of consistent and sensible values of vacancy formation energy (see § 4.3.3). Analyses in which the asymptotic values of annihilation rate,

λ_1, λ_2, at low and high temperatures (fig. 37) are applied to eqn. (82) to provide a fit to the spectra at intermediate temperatures (Dave *et al.* 1971, Snead *et al.* 1972) provide statistically convincing results.

Unconstrained, model-independent analyses of multi-component lifetime spectra require data considerably more accurate than that usually obtainable. Recent investigations by Crisp *et al.* (1973) are moderately successful. Modifications to a conventional lifetime spectrometer allowed prolonged experiments (~ 10 days) and the accumulation of the order of 5×10^5 counts per channel at the spectrum peak. Multi-component analysis established the excellence (χ^2 per degree of freedom) of a three (two + source) component fit for the spectra for carefully annealed polycrystalline indium samples (table 7). At 405°K, the small intensity and fast decay of the shortest-lived component results in unavoidably large errors in the fitting procedure.

Table 7. Component intensities and rates obtained from an unconstrained analysis of lifetime spectra for well-annealed polycrystalline In. The data in the last two columns are obtained from a trapping-model analysis of the data in columns 1–4 (Crisp *et al.* 1973)

Sample temp.	I_1 (%)	Γ_1^{-1} (psec)	I_2 (%)	Γ_2^{-1} (psec)	λ_1^{-1} (psec)	K_{12} (nsec)$^{-1}$
83	71.6 ± 4.7	157 ± 5	28.4 ± 7.2	275 ± 22	179 ± 9	0.83 ± 0.14
170	71.0 ± 3.1	159 ± 4	29.0 ± 5.2	283 ± 13	181 ± 4	0.81 ± 0.10
298	65.2 ± 5.3	152 ± 7	34.8 ± 5.1	266 ± 19	188 ± 12	0.90 ± 0.17
359	32.4 ± 1.5	123 ± 8	67.6 ± 4.3	254 ± 7	190 ± 8	2.86 ± 0.34
405	7.4 ± 4.6	106 ± 56	92.6 ± 5.1	245 ± 5	205 ± 27	8 ± 7

At low temperatures a second longer-lived component of constant intensity and lifetime is associated with residual defects—possibly grain boundaries. A similar intensity of temperature-independent component has been observed for well-annealed samples of aluminium (Snead *et al.* 1972). The increase in I_2 at higher temperatures (table 7) is consistent with the predictions of a trapping model and the increasing concentration of vacancies. At these temperatures I_2 presumably represents the sum of two unresolved components of similar rate. Although more complex patterns of transitions cannot be entirely ruled out, the simple (no escape) trapping models would appear to provide a realistic description of vacancy and other defect-trapping phenomena in this metal.

(ii) *Mechanisms of positron trapping*

Application of the two-state trapping model to the admittedly poorly defined parameters listed in table 7 yields relative trapping rates, $K_{1v} = \nu_{1v} C_v$, at 298 and 359°K in reasonable agreement with relative vacancy concentrations obtained from the data of table 10. The measurement of absolute trap concentrations demands a theory for ν_{1v}. An agreed model has not been established. Naive considerations suggest (Connors and West 1969, Snead *et al.* 1971, Cotterill, Petersen, Trumpy and Träff 1972)

$$K_{1v} = \sigma_v \bar{v} \rho A \exp \left(- E_f / k_B T \right), \tag{88}$$

where σ_v is the cross section for positron-vacancy trapping, $\bar{v} = (8k_B T / \pi m)^{1/2}$ is the speed of the thermalized positron, ρ is the atomic density, and $A \exp \left(- E_f / k_B T \right)$ is the equilibrium vacancy concentration.

Values for the product $\nu_{1v} A$ have been reported for several metals and usually fall in the range 10^{14}–10^{17} sec^{-1}. For aluminium, consistent values for $\nu_{1v} \sim 2 \times 10^{15}$ sec^{-1} have been obtained from both equilibrium (McKee, Jost and MacKenzie 1972, McKee, Triftshäuser and Stewart 1972 and quenching studies (Cotterill, Petersen, Trumpy and Traff 1972). Some workers (Connors and West 1969, Hautojärvi *et al.* 1970, Snead *et al.* 1971, Cotterill, Petersen, Trumpy and Traff 1972) have used eqn. (88) to deduce values for σ_v which are plausibly of the order of 10 to 100 Å^2. However, the experimental uncertainties in these values are large and considerably more data will be required to establish the validity of the more sophisticated theories of the trapping process.

A zero-temperature quantum-mechanical calculation of trapping probability has been made by Hodges (1970) who obtains $\nu_{1v} \approx 10^{15}$ sec^{-1} for vacancies in aluminium. The temperature dependence of ν_{1v} is more controversial. Positron trapping by vacancies will clearly depend both on the positron state and dynamic properties in the normal regions of the crystal and its fundamental interaction with vacancy centres.

Brandt and Waung (1971) have suggested a picture in which the positrons interact weakly with the lattice and move as free particles of effective mass m_+^*, velocity v_+, and wavelength $\lambda = \hbar / m_+^* v_+ \gg R$, where R is the effective vacancy radius. The positrons interact with the vacancies with a trapping cross section σ_v which depends on λ and the positron wavenumber inside (K_+) and outside (k_+) the vacancy. To lowest order in k_+,

$$\sigma_v = 4\pi \lambda^2 \left(\frac{k_+}{K_+} \right)$$

A T^0 dependence for ν_{1v} would seem to be implied. Brandt and Waung claim a capture cross section similar to those observed in metals but details are not given.

Seeger (1972 a, b, 1973) adopts a rather different approach. Analogy with the problem of electron mobility in non-degenerate semiconductors (Smith 1969) suggests a positron mobility

$$\mu = \frac{2^{3/2} \pi^{1/2} \rho C_l^2 \hbar^4}{3 m_+^{5/2} \epsilon_d^2 (k_B T)^{3/2}} \tag{89}$$

which is used with the Nernst–Einstein relation $D = k_B T \mu$ to define a positron diffusion coefficient D. The trapping rate,

$$\nu_{1v} = 4\pi D r_0 \Omega^{-1}, \tag{90}$$

where Ω is the volume of a unit cell, and r_0 is a temperature-independent capture radius. In eqn. (89), ρ is the density, C_l is the longitudinal sound velocity, and m_+ is the positron mass. The deformation potential constant, $\epsilon_d = \delta\epsilon/\Theta$, where $\delta\epsilon$ is the change in positron ground-state energy which results from a uniform dilatation Θ. A $T^{-1/2}$ dependence for the trapping rate is thus predicted. A numerical estimate of ν_{1v} based on eqns. (89) and (90) is in reasonable agreement with experiment if a positron effective mass of a few free-electron masses is assumed (Seeger 1972 b).

A precise theoretical justification for the adoption of a particular T dependence for ν_{1v} is hard to see at present. A $T^{-1/2}$ dependent mobility is more easily justified for positrons in semiconductors (Bergerson and McMullen 1971). In metals, a somewhat weaker positron–lattice coupling should result from conduction electron screening. Effective mass calculations for alkali metals (Mikeska 1970) suggest that electrons and phonons provide similar contributions to the positron scattering. In heavier metals, where the angular distributions show appreciable core annihilation the positron–phonon coupling should be more important. A more quantitative analysis must be left to the theorists.

Experimental evidence as to the temperature dependence of ν_{1v} is scarce. Seeger (1972 b, 1973) cites the quenching study of Connors *et al.* (1970) but reference to fig. 38 will show that this cannot be justified. Furthermore, the algebraic analysis in Seeger (1972 b) would appear to be in error (Connors and Bowler 1973). Explicit studies of the temperature dependence of trapping by ' frozen in ' vacancies in quenched samples of higher melting point metals are now being made. Preliminary reports of the temperature dependence of ν_{1v} in gold vary from T^0 (T. M. Hall, private communication) to T^1 (B. T. A. McKee, private communication) and thus are as divergent as the theoretical predictions.

(iii) *Trapped positron states*

The data of table 7 suggest a slight but plausible decrease in λ_1 with increase in temperature or volume. The very much smaller annihilation rate for trapped positron states can be qualitatively explained from a comparison with angular correlation changes.

First we consider vacancy traps in which the absence of a positive ion provides an attractive potential for positrons and a repulsive potential for conduction electrons. A localization of the positron around a vacancy site will result in a decrease in the electron density seen by the positron. The decrease in core electron density will reflect the degree of localization of the positron state (Connors *et al.* 1971, Dave *et al.* 1971). A description of the decrease in conduction electron density at the positron must be more complex. The repulsive vacancy–conduction electron interaction will tend to decrease the annihilation rate but this interaction will be compensated in part by the attractive field of the trapped positron. Electron–positron correlation enhancement factors (§ 2.42) for trapped positrons may be different from those for a free positron (Hodges 1970, MacKenzie, Craig and McKee 1971). The decreased contribution of core electrons to the screening of the positron–conduction electron interaction may be important in some metals (West 1971).

In table 8 we have used the analysis described in § 2.4.3 to deduce the partial core and conduction electron annihilation rates in several metals at (a) low temperatures where vacancy trapping is small, and (b) close to the melting point where vacancy trapping is probably complete. Subject to the limitations of such an analysis (see § 3.2) the following observations may be made. In the divalent metals zinc and cadmium the observed change in total annihilation rate is almost entirely due to the decrease in core electron rate (Connors et al. 1971). This fact alone implies a considerable potential energy advantage for a positron trapped in a vacancy. An apparent balance between the various factors affecting the conduction electron rate is upset as the valence increases. In trivalent indium and aluminium the more severe positron localization and greater repulsion of conduction electrons, implied by the larger effective vacancy charge, are now apparent in the decrease in conduction electron annihilation rate.

Table 8. Valence electron annihilation rates in low temperature (no trapping) and high temperature (saturated trapping) metals. The angular correlation data are taken from Kusmiss (1965). The annihilation rates have been obtained from the curves of Mackenzie et al. (1967). For any entry the significance of the last decimal place is doubtful

Metal	Valence electron contrib. to ang. dist.		Total annihilation rate		Valence electron rate	
	Low T	High T	Low T	High T	Low T	High T
Zn	0·52	0·68	5·7	4·5	3·0	3·0
Cd	0·61	0·69	5·1	4·3	3·1	3·0
In	0·57	0·68	5·1	4·0	2·9	2·7
Al	0·81	0·91	6·0	4·1	4·9	3·7

The large decrease in core annihilation rate for all four metals suggests strong positron localization and thus supports the use of the simple (zero escape) trapping models. A more quantitative assessment of positron-vacancy binding or positron localization, ideally, should take into account not only the interaction with the background ion system, but also the very mportant electron–electron and positron–electron correlation effects.

Hodges (1970) adopts a neutral pseudoatom approach (Ziman 1967) in which the positron and ions each with their screening clouds of electrons interact through a suitable self-consistent potential. A square-well approximation to the vacancy potential provides the positron binding energies listed in table 9. The binding energies are measured relative to the positron ground–state energy in the perfect lattice and, in the main, their relative magnitudes correlate well with the experimental observations of positron trapping. However, the sensitivity of the results to the partial inclusion of correlation effects leaves some doubt as to the significance of absolute values which, contrary to experimental evidence, suggest positron trapping in all the metals investigated. Seeger (1973) has given qualitative explanations for the non-observation of trapping in some metals.

The relatively small contribution of core electrons to the angular distributions for aluminium has provided the spur for a more complete description of the positron and electron states in the vacancy region. Arponen *et al.* (1973) have treated this problem in some detail. A square-well approximation to the ionic part of the vacancy potential is modified by the potential that results from the redistribution of conduction electrons around the vacancy. The modifications to the conduction electron density are deduced from a Thomas–Fermi calculation. The calculated positron binding energy of 3·8 eV is intermediate between the values obtained by Hodges (table 9).

A calculation of the consequent positron lifetime or angular distribution is less straightforward. The vacancy-induced changes in angular distribution include significant narrowing and smearing of the normally parabolic conduction electron component. Any change in the non-local conduction electron momentum distribution, at even the maximum vacancy concentration attainable in a solid, will be relatively minor (Stern 1968). However, an appreciable localization of the positron wavefunction can be expected to produce rather stronger, local effects in the pair distribution because of its dependence on the wavefunction product (eqn. 15).

Arponen *et al.* (1973) deal with these problems in the following way. A local density approximation (Brandt 1967) to the effects of enhancement (§ 2.4.2) assumes that at any point, r where the electron density is $\rho_e(r)$ the total annihilation rate is given by the product of the positron density $|\psi_+(r)|^2$ and the annihilation rate for a homogeneous electron gas of the same density ρ_e. A calculation involving the computed positron and electron densities is in good agreement with the experimental results for saturated positron trapping (Cotterill, Petersen, Trumpy and Träff 1972, McKee, Jost and MacKenzie 1972). The calculation of an angular distribution uses an approximate version of eqn. (11) which takes partial account of the phase or non-local aspects of the problem. The result of this ' mixed density approximation ' (the Arponen *et al.* term) is shown in fig. 39. (The considerably poorer result of an approximation which neglects the phase effects is also shown.) The large angle tail is due primarily to the modifications to the positron momentum distribution which arise from the localization. Thus this smearing provides an additional useful measure of the localization or strength of positron trapping (Connors *et al.* 1971, Dave *et al.* 1971, Hautojärvi 1972).

Table 9. Positron-vacancy binding energies (Ryd) calculated in (1) a self-consistent Hartree approximation and (2) including some of the effects of electron–electron correlations (Hodges 1970)

	Li	Na	K	Rb	Cs	Mg	Zn	Cd	Hg	Al	Ga	In	Tl	Sn	Pb
$E_b^{(1)}$	0·05	0·06	0·07	0·08	0·07	0·18	0·19	0·17	0·11	0·28	0·20	0·22	0·17	0·24	0·22
$E_b^{(2)}$	0·01	0·01	0·01	0·01	0·01	0·07	0·08	0·06	0·02	0·15	0·08	0·10	0·06	0·10	0·09

Table 10. Vacancy formation energies obtained from positron trapping studies

E_f (eV)					Method	Reference
Zn	Cd	Al	In	Pb		
0·50 ± 0·02	0·39 ± 0·03	0·67 ± 0·04	0·54 ± 0·04	0·50 ± 0·03	Angular correlation h-parameter	McKee, Trifthäuser and Stewart (1972)
	0·52 ± 0·05				Angular correlation h-parameter	Connors et al. (1971)
0·51 ± 0·025	0·42 ± 0·02	0·68 ± 0·03	0·46 ± 0·03	0·49 ± 0·03	Centroid shift t	McKee, private communication (1973)
		0·64 ± 0·02			Lifetime spectrum analysis	Dave et al. (1971)
		0·62 ± 0·02			Lifetime spectrum analysis	Snead et al. (1972)

Fig. 39

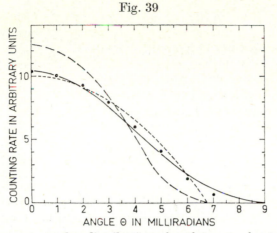

Conduction electron angular distributions for the case of saturated vacancy
trapping in Al. Experimental points from Kusmiss and Stewart (1967 a).
The calculated curves correspond to a mixed density approximation (solid
line), a local density approximation (broken line), and the free electron
parabola (dotted line) (Arponen *et al.* 1973).

The vacancy-induced modifications to the conduction electron part of
angular distributions which exhibit larger core contributions are qualitatively similar to those described above. However, a quantitative assessment of these changes, including the smearing, may be complicated by
the core distribution. A constancy of the core distribution shape at large
angles does not exclude the possibility of significant changes at low angles.
The discussion of § 3.2 is relevant.

The nature of the defects which produce the rather more pronounced
narrowing of angular distributions (Kusmiss *et al.* 1970, Snead *et al.* 1971)
and changes in lifetime (Cotterill, Petersen, Trumpy and Träff 1972) in
deformed metals is not entirely clear. Atomistic and continuum models
for dislocations in aluminium have been applied to the positron trapping
problem (Hautojärvi 1972, Martin and Paetsch 1972). Continuum models
of dislocations result in positron-dislocation binding energies < 0.1 eV,
too small to explain the observed changes in lifetime or angular correlation.
The similarlity of dislocation and vacancy-induced changes suggest that
the positron 'sees' a similar environment in any defect and positron
trapping by dislocation cores has been proposed (Cotterill, Petersen,
Trumpy and Träff 1972, Hautojärvi 1972, Martin and Paetsch 1972). A
theoretical, atomistic treatment of this problem will clearly be complex
and may require the consideration of positron states of unusual symmetry
(Martin and Paetsch 1972).

4.3.3. *Applications*

Positron annihilation measurements of vacancy formation energy are a
particularly attractive proposition since they can be made at temperatures

where the contribution of divacancies to the effective formation energy is negligible (McKee, Triftshäuser and Stewart 1972, Seeger 1973). The results presented in table 10 include some privately communicated corrections. Seeger (1973) has presented a point-defect specialist's view of the techniques and results. Thus a short account, as it were, from the other side of the fence would seem to be in order.

Assuming the two-state trapping model (eqn. (82)), we deduce from eqns. (84) and (85) that, at any temperature T,

$$F = \frac{F_1\lambda_1 + F_v K_{1v}}{\lambda_1 + K_{1v}}. \tag{91}$$

Here F is a mean positron lifetime (t) or height of a normalized angular distribution (h) and F_1 and F_v are, respectively, the low temperature (zero trapping) and high temperature (saturated trapping) values. The temperature dependence of F enters through K_{1v} for which, in view of our earlier observations (§ 4.3.2), we write

$$K_{1v} = \nu_{1v}c_v = A \exp\left(-E_f/k_B T\right). \tag{92}$$

A is the product of ν_{1v} and the usual entropy factor $\exp\left(S_v/k_B\right)$ (Seeger 1973). Any temperature dependence of ν_{1v} of the forms discussed in § 4.3.2 will be relatively minor compared with that of the exponential factor and, for the present, will be neglected. The applicability of eqns. (92) and (91) (Bergersen and Stott 1969, Connors and West 1969) to the lifetime results of MacKenzie et al. (1967) can be seen in the curves of fig. 37. A more revealing presentation (Connors et al. 1971, McKee, Triftshaüser and Stewart 1972) can be obtained from the Arrhenius plot (fig. 40) suggested by the easily deduced equation

$$\ln\left[\frac{F - F_1}{F_v - F}\right] = -\frac{E_f}{k_B T} + \ln\left[\frac{A}{\lambda_1}\right]. \tag{93}$$

The results of table 10 derive from a variety of analyses. McKee et al. (McKee, Triftshäuser and Stewart 1972, McKee, private communication) used the variables F_f, F_v, E_f and A/λ_1 as adjustable parameters in fitting their data to eqn. (91) (with K_{1v} as in eqn. (92)). It is difficult to assess the reliability of this approach but the good agreement (indium excepted) between the results of angular correlation and lifetime studies gives some cause for comfort and an agreeable confirmation of the analysis of § 4.1.2. Connors et al. (1971) applied additional contraints by fitting angular distribution data to eqn. (93) with $h_1(T)$ and h_2 values deduced from low and high temperature measurements.

Similar constraints were applied by Dave et al. (1971) in their treatment of the aluminium lifetime results of McKee, Jost and MacKenzie (1972). Here, experimentally deduced asymptotic values of annihilation rate (λ_1, λ_v) were applied with the trapping model, eqn. (82), directly to the measured spectra. The linearity of an Arrhenius plot of

$$\ln\left(\frac{I_2}{I_1}\right) = \ln\left(\frac{K_{1v}}{\lambda_1 - \lambda_v}\right) \text{ versus } \frac{1}{T}$$

(eqns. (82) and (92)) gives apparent support for both the deduced value of E_f and the applicability of the simple *two*-state trapping model. Whether or not the deduced value for E_f is in any way better than that which would be obtained from a similarly constrained fit to eqn. (91) is not clear to this author. However, the intrinsic value of this type of approach is more clearly demonstrated by the work of Snead *et al.* (1972) whose almost identical analysis of their own aluminium lifetime spectra disclosed a non-linear Arrhenius plot which was attributed to an additional temperature-independent defect component. A modified analysis which took account of this component provided better fits to the esperimental data but of necessity contained rather more adjustable parameters.

Fig. 40

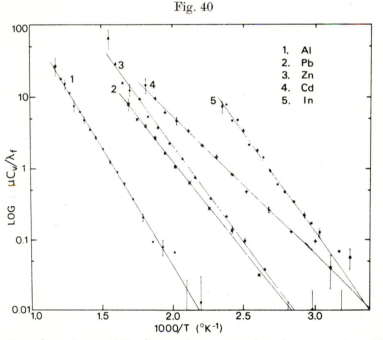

Arrhenius plots (eqn. (93)) for positron trapping by vacancies (McKee, Trifthäuser and Stewart 1972).

McKee, Jost and MacKenzie (1972) analysed their data in terms of both trapping (82) and enhancement (79) models.

A critical assessment of the more technical aspects of any analysis is best left to its user. The writer's own experience would suggest the superiority of those analyses (when possible) in which the asymptotic values of annihilation rate (λ_1, λ_v) are used in fitting the experimental results to any equation. However, in some cases the 'best' method of analysis may depend on the need for modifications to the basic trapping model (Snead *et al.* 1972).

Significant differences between E_f values (table 10) obtained by different workers suggest the need for a more critical examination of the common, but not always necessary, assumptions applied to the trapping model.

Comparatively little investigation has been made of the temperature dependence of the parameters λ_1, h_1, λ_v, h_v. Some temperature dependence of λ_1 and h_1 is made inevitable by thermal expansion. Phonon effects, which as we have seen can be invoked in the calculation of trapping rates (Seeger 1972 a), can also be expected to produce an even stronger decrease in λ_1 with increase in temperature. Some increase in h_1 could also be caused by suppression of HMC (see § 3.1.2 and Brandt and Reinheimer (1971). A quantitative theory for all these effects would be difficult but experimental investigation is not impossible.

The data of table 7 indicate a small but significant temperature dependence for λ_1. A similar variation for h_1 in low temperature cadmium has been reported by Connors *et al.* (1971) and its inclusion in the subsequent calculation of E_f (table 10) accounts for some 0·3 eV of the comparatively large value obtained by these workers.

A temperature-dependent decrease in λ_1 could be accompanied by an increase in λ_v. The complication of melting usually precludes a measure of corresponding changes in h_v but continued improvement in lifetime measurement techniques may allow for the temperature dependence of λ_v to be exposed by spectrum analysis. The neglect of significant variations in λ_1 and λ_v of the type suggested would be additive in its effect on E_f and could produce significant error.

Other inadequacies in the trapping model are less likely to have had an appreciable effect on the results of table 10. A severe localization of the positron around a vacancy is suggested (§ 4.3.2) by both the calculations of binding energy (Hodges 1970) and estimates of smearing and other effects in angular distribution (Connors *et al.* 1971, Dave *et al.* 1971, Hautojärvi 1972). Nevertheless, the absolute values for binding energies may be in error and the reliability of any analysis of an angular distribution will always depend on our questionable ability to define precisely the core annihilation. Even the existence of a multi-component lifetime spectrum does not entirely rule out escape (Connors *et al.* 1971) and a variety of different types of spectra can provide essentially similar behaviour in highly averaged quantities like t and h.

Appreciable escape will certainly be more probable in the case of high melting-point metals and may be the reason for certain anomalies in the temperature dependence of doppler broadening (MacKenzie, Gingerich and Kim 1971) and angular distributions (Sueoka 1969) for copper. However, divacancy trapping (Seeger 1973), or electronic structure anisotropy effects (Sueoka and Ishihara 1972) may play a part.

The results presented in table 7 and those of Snead *et al.* (1972) suggest the presence of other effective positron traps in well-annealed specimens. The rather limited evidence available suggests the additional trapping is temperature independent but the consequent complication of data analysis (Snead *et al.* 1972) points to the need for more detailed examination of other lifetime spectra.

In the light of the discussion of previous paragraphs we would suggest that the application of the relatively minor corrections that apply to a $T^{-1/2}$ dependence of trapping rate (Seeger 1973) is probably premature.

Positron annihilation measurements of vacancy formation energies have now been reported for Zn, Cd, Al, In, and Pb (table 10), Cu, α, β, and γ-brass (MacKenzie *et al.* 1971b). Any objection to the E_f values deduced for aluminium would be hard to sustain. The values for other metals in table 10 are worthy of re-examination for possible complications of the sort described above. This remark is even more relevant to the case of measurements in high melting point metals (MacKenzie, Gingerich and Kim 1971).

Theoretical inadequacies have inhibited quantitative measurements of dislocation trapping effects. Nevertheless we would point out the remarkable ability of positron studies to reflect the history or monitor the state of defected material. For example, lifetime (Hautojärvi *et al.* 1970) or angular correlation studies (Connors 1971, unpublished), when analysed in terms of the trapping models, are in close agreement as to the state of defected samples (fig. 41). Lineshape measurements (MacKenzie *et al.* 1970) again provide a compelling and quantifiable relation between positron trapping and deformation. More important perhaps, a close correlation between positron trapping and hardness suggests a valuable non-destructive technological application. Positron lifetimes in shock-deformed polycrystalline nickel samples have been measured by Holt *et al.* (1970). Again a quantifiable relation between positron lifetime and sample history is easily shown.

Particularly timely are the investigations of positron trapping by the large voids induced by fast neutron irradiation of metals. The formation of such voids and the concomitant swelling, and possible rupture, provides a severe problem for the designers of fast power reactors. Neutron irradiated molybdenum samples containing voids of 30–40 Å diameter have been investigated (Mogensen *et al.* 1972, Cotterill, MacKenzie, Smedskjaer, Trumpy and Träff 1972) and provide effects in angular correlation, lineshape, and lifetime, similar to, but of several times the magnitude of those already attributed to vacancies or dislocations.

Positron annihilation measurements undoubtedly have considerable potential in studies of defected material. Lifetime measurements can, in principle, supply quantitative information about defect concentrations. Angular correlation studies should, in time, be able to provide a picture of electron states in defect regions. The rate of data collection in annihilation lineshape measurements suggests a considerable potential for investigations of rapid annealing or recovery processes.

Fig. 41

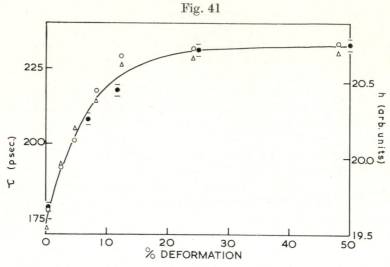

A comparison of mean lifetime and angular correlation changes after deformation of Al.   Full circles :   h-parameter (Connors, unpublished).   Open
circles and triangles :   t (Hautojärvi et al. 1970).

There are, of course, limitations.   A significant positron sensitivity to
defects has been observed in relatively few of the more common metals.
The apparent absence of trapping phenomena in some metals may be, in
part, a result of imprecise measurements.   Nevertheless, should more
subtle enhancement effects eventually be resolved a quantitative analysis
of the underlying phenomena will certainly be more difficult.   In even
those materials where positron trapping is severe the observation of
significant effects in measurable parameters will be restricted to a finite
range of defect concentrations.   In those cases where the simple trapping
models (eqn. (80)) apply, a simple criterion, $0 \cdot 01 \lambda_1 \lesssim \nu_{1j} c_j \lesssim 100 \lambda_1$ and
$\nu_{1j} \sim 10^{15}$ sec^{-1} (§ 4.3.2) can serve as a guide to sensible experiments.

4.3.4. Alloys

The interpretation of defect-induced changes in positron annihilation
parameters in impure metals (Sedov et al. 1970, Snead et al. 1972, Tanigawa
et al. 1972) and alloys (McKee et al. 1967, MacKenzie, Gingerich and Kim
1971) necessarily must be more complex.

Trapping-model analyses of low impurity systems, in terms of vacancy–
impurity binding energies (Snead et al. 1972) make strong demands on
measurement technique and require many assumptions (Snead et al. 1972,
Seeger 1973).

A marked suppression of defect-trapping phenomena at higher impurity
concentrations (Tanigawa et al. 1972) can be attributed to other effects.
Vacancy–impurity binding effects may be accompanied by a reduced
attraction of positrons to vacancy–impurity pairs (Tanigawa et al. 1972,

Seeger 1973). The atomic disorder may significantly affect the positron mobility.

In alloys of dissimilar metals, positron-impurity or impurity cluster trapping can be important (Sedov *et al.* 1970, Kubia *et al.* 1971). At large concentrations, an equation of the form, $K = vc$ (eqn. (83)) is unlikely to be appropriate and a more complex concentration dependence of impurity trapping or enhancement effects can be expected.

Such problems may be important in both defect (MacKenzie, Gingerich and Kim 1971) and electronic structure (see § 3.3.2) measurements in alloys and experimental and theoretical effort in this area would seem to be worth while.

4.3.5. *Liquid metals*

Significant changes in angular correlation at melting have been reported for tin, gallium, copper, mercury and bismuth (Gustafson and Mackintosh 1963, Gustafson *et al.* 1963, West *et al.* 1967, Kusmiss and Stewart 1967 a, Mogensen and Trumpy 1969, Itoh *et al.* 1972). The changes, which are most pronounced in mercury (fig. 42) are qualitatively similar to those arising from defect trapping in solid metals (fig. 36) and generally occur in those metals where vacancy effects are theoretically likely (Seeger 1973) but surprisingly not observed in the solid phases. A logical association of these effects with positron 'trapping' in low density regions of the disordered liquids is supported by observations of appropriately large changes in positron lifetime in mercury and gallium (fig. 43).

Fig. 42

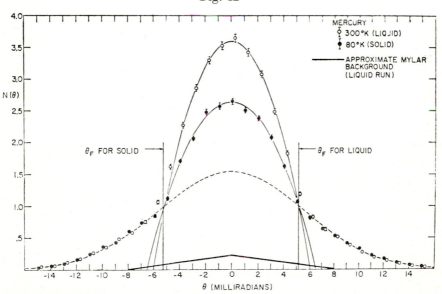

Long slit angular distributions in solid and liquid Hg (Gustafson *et al.* 1963).

Fig. 43

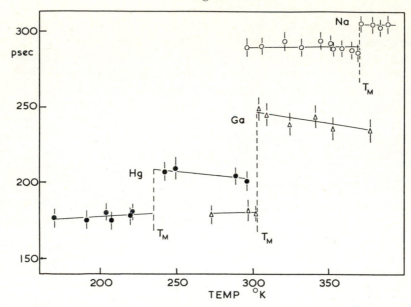

Positron lifetimes in solid and liquid Na, Ga (Brandt and Waung 1968), and Hg
(West *et al.* 1973).   The solid lines are merely visual fits to the data.

The analogy with vacancy trapping in solid metals is made clearer by a
more general survey.   Metals that demonstrate large changes in angular
correlation and lifetime on heating in the solid usually show much smaller
additional changes on melting (table 6).   The alkali metals show little or
no change with either temperature or melting (Kusmiss and Stewart
1967 a, Arias-Limonta and Varlashkin 1970 a) other than the smearing
already associated with positron effective mass (§ 2.4.4).

Early attempts to interpret angular distributions for liquid metals made
use of the striking similarity between the large angle parts of the distribution for each phase (fig. 42).   The smearing of the conduction electron
part of the angular distribution for the liquid metal was then attributed
(Gustafson *et al.* 1963, West *et al.* 1967) to the effect of disorder on the
conduction electron states (Ballentine 1966).

Our current understanding of positron defect-trapping phenomena in
solid metals provides a somewhat different picture.   The smearing and
narrowing (West and Cusack 1967, Mogensen and Trumpy 1969) of the
conduction electron distribution is as observed for vacancy trapping and
may be similarly interpreted in terms of ' positron localization ' and ' local
density ' effects (§ 4.3.2).   A quantitative analysis again depends on our
ability to construct a realistic core component (§ 3.2).   Early investigations indicated a close similarity between the core distributions in each
phase but subsequent, more precise measurements (Mogensen and Trumpy
1969) show significant differences which again may be attributed to the

structural disorder (Brandt and Reinheimer 1971). These differences are likely to be even larger at low angles and could contribute to the apparent narrowing of the total angular distribution.

The nature of the underlying positron traps is of interest. An investigation of positron traps in solid metals is made somewhat easier by the comparative persistence of these traps relative to positron lifetimes (§ 4.1.1) In liquid metals, a size distribution of positron traps, the transient nature of the traps and the possibility of the positron creating or stabilizing traps must all be considered. Spectral analyses of positron lifetime spectra in solid and liquid mercury (West *et al.* 1973) are relevant. Measurements employing both homogenized Hg–^{22}Na amalgam and conventional sandwich sources indicate a single-component lifetime in the solid phase. Multi-component (1→3) analyses of results for liquid mercury are consistently unsuccessful and indicate a more complex type of spectrum possibly of the form suggested by eqn. (74). Investigations over a wider range of temperatures would be of value.

The somewhat smaller effect of disorder on positron states in liquid alkali metals has been assessed by Nakai *et al.* (1973) with a Green's function calculation. The results are consistent with our general observations and suggest that modifications to the positron momentum distribution produce the bulk of the smearing in angular distribution.

4.4. *Powders, voids, surface, and other phenomena*

Positron annihilation results for single-crystal ionic solids are often indistinguishable from those from the polycrystals (Bertolaccini *et al.* 1971). Some metallic oxides are hard to obtain in other than microcrystalline form and positron studies often make use of compacted powder specimens which, not surprisingly, provide more complex results.

Lifetime spectra for oxide powders usually consist of three components of which the longest lived has a lifetime (10→100 nsec) approaching that of the natural decay of the orthopositronium atom (Paulin and Ambrosino 1967, Sen and Patro 1969). Further, angular distributions for powdered oxides exhibit a narrow parapositronium peak (Kusmiss and Stewart 1967 b) which is enhanced by oxygen quenching (Mills 1968, Sen and Patro 1969).

The two shorter-lived lifetime components ($\tau_1 \sim 0.5$ nsec, $\tau_2 \sim 2$ nsec) are typical of the spectra for the bulk material in those cases where it can be investigated. Various correlations between these lifetimes and crystal properties have been suggested (Sen and Patro 1969, Bertolaccini *et al.* 1971, Chiba *et al.* 1971) but as yet no really consistent picture has emerged.

A plausible, qualitative interpretation of the spectra for powder samples (Brandt and Paulin 1968) identifies the fastest component (I_1, τ_1) with parapositronium and free positron decay and the intermediate component (I_2, τ_2) with the pick-off annihilation of orthopositronium atoms. The third longest lived component has an intensity, I_3, which grows rapidly, at the expense of I_2 with decrease in particle size.

This effect can be quantitatively explained by a theory (Brandt and Paulin 1968, 1972) which assumes that the orthopositronium atoms are formed in the homogeneous interiors of particle grains and subsequently migrate by random walk until they annihilate or escape into the regions or voids between the grains. Positronium annihilation within the solids where the free volume (§ 2.2.3) is small results in comparatively swift orthopositronium decay ($\tau_p \sim \tau_2$) and a zero point energy broadened parapositronium peak in angular distribution (Steldt and Varlashkin 1972). On the other hand, orthopositronium atoms that escape into the large free volume regions between the particles annihilate more slowly (τ_3) and give rise to a narrower parapositronium peak.

An exact theory is complex (Brandt and Paulin, 1972) but provides a relation between I_3 and a parameter β :

$$\beta = (D\tau_p)^{1/2}/R,$$

where D is the positron diffusion constant, τ_p is the orthopositronium pick-off rate in the particles and R is the radius of a (spherical) particle.

An evaluation of experimental data (particle size ~ 100 Å) provide positron diffusion constants (Brandt and Paulin 1972)

$$\left.\begin{array}{l} D(SiO_2) = 1\cdot45 \pm 0\cdot15 \\ D(Al_2O_3) = 5\cdot5 \pm 1 \\ D(MgO) = 25 \pm 15 \end{array}\right\} \times 10^{-5} \text{ cm}^2 \text{ sec}^{-1}.$$

Measurements of the temperature dependence of I_3 disclose a variation consistent with a temperature dependent diffusion constant $D \propto \exp(-\theta/T)$ (Brandt and Paulin 1968).

Angular correlation studies in these materials produce dramatic evidence of this positronium diffusion phenomenon (Steldt and Varlashkin 1972). Under normal circumstances relatively few of the parapositronium atoms will escape from the solids. In SiO_2, the orthopositronium pick-off lifetime ($\tau_2 \sim 2$ nsec) is some sixteen times the parapositronium lifetime ($t_{para} = 0\cdot125$ nsec). In consequence, an orthopositronium atom will migrate a distance $(2D\tau)^{1/2}$ some four times that for a para-atom (Brandt and Paulin 1968). This and the usual statistical factor (§ 1.1) suggests that only one in sixteen of the positronium atoms that leave the solids will be of the para-variety.

Steldt and Varlashkin (1972) have studied angular correlation in powdered SiO_2 and have systematically examined the effect of particle size (50, 120, 400 Å), temperature, environment, surface condition, and density, on the intensity and form of the narrow component. The utilization of an 8 kG magnetic field to focus the positrons onto the sample also provided magnetic quenching of some of the orthopositronium atoms. The parapositronium contributions to each distribution were assessed by comparison with the angular distribution for crystalline quartz (fig. 44).

Fig. 44

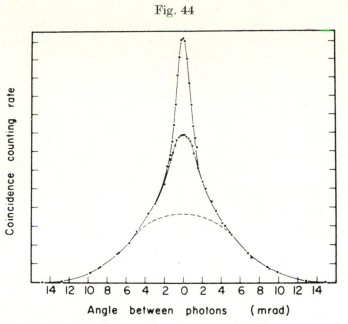

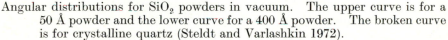

Angular distributions for SiO_2 powders in vacuum. The upper curve is for a 50 Å powder and the lower curve for a 400 Å powder. The broken curve is for crystalline quartz (Steldt and Varlashkin 1972).

The parapositronium peak for the 400 Å powders is broad and consistent with a significant contribution from parapositronium decay in small free volume regions. The results for 120 and 50 Å powders show an increased intensity of parapositronium component of the narrow width appropriate to annihilation in the large free volume voided regions. The diffusion-controlled contribution to the narrow peak is clearly indicated by a temperature-dependent intensity in powders under vacuum. The parapositronium intensity and its temperature dependence is more marked for samples in oxygen atmosphere consistent with strong oxygen quenching (§ 2.1) of orthopositronium atoms once they reach the voids.

Of more far-reaching significance is the observation of an enhanced intensity of narrow component in samples which have previously been baked at elevated temperatures to remove surface absorbed water. The possibility that the positronium atoms in the voids are strongly associated with the enclosing surface (Steldt and Varlashkin 1972) has also been recognized by other workers (Brandt and Paulin 1968, Gainotti and Ghezzi 1970, Chuang *et al.* 1971).

Chuang *et al.* (1971) have noted that the comparatively slight dependence of τ_3 on pore size (25→136 Å) in silica gels cannot be explained by conventional free volume theories of pick-off annihilation. This, and a marked sensitivity of τ_3 to molecules absorbed on the pore surface, are used to suggest the absorption of positronium atoms on the surface.

The discussion of positronium affinity in § 2.3 provides good reasons for supposing that positrons will find it energetically advantageous to escape ionic solids. However, a discussion of surface states must also take account of the weaker Van der Waals forces between the positronium and surface atoms. Chuang *et al.* (1971) represent the surface potential by a simple square model. Their study suggests that surface states exist but are only weakly bound.

The possibility of positron states on metallic surfaces has been proposed, albeit with little conviction, by Mogensen *et al.* (1972) as part of their interpretation of positron trapping studies of voided molybdenum (see § 4.3.3). The theoretical plausibility of positronium formation in the large ($10 \rightarrow 100$ Å) voids is not borne out by magnetic quenching studies despite the large increases in positron lifetime and pronounced narrowing of angular distribution or annihilation line shape introduced by the voids (Mogensen *et al.* 1972, Cotterill, MacKenzie, Smedskjaer, Trumpy and Träff 1972).

Dekhtyar *et al.* (1972) suggest that positron surface states provide the origin for anisotropies and structure in their angular distributions for single-crystal molybdenum samples. Theoretical fuel for the discussion of positron escape from solids has been provided in response to phenomena which indicate the possibility of negative positron work functions for some metals. When high energy positrons are thermalized in a moderator (usually mica) coated with a thin metallic film of aluminium or gold a small number of positrons leave the free metal surface with an energy distribution that peaks at a few eV (Pendyala *et al.* 1971, Costello *et al.* 1972).

The positron work function can be regarded as being made up of three contributions.

1. The positron correlation energy, E_{corr} which arises from the attractive electron–positron interaction (§ 2.4.2).
2. The zero point or ground state energy E_0, which depends on the repulsive positron–ion interaction (§ 3.1.1).
3. The work done, D, in surmounting the potential barrier of the surface dipole layer which is caused by the relaxation of surface electron states.

The most recent assessment of all three effects has been made by Hodges and Stott (1973). The calculation of positron–electron correlation energy was based on the dielectric screening theory of Singwi *et al.* (1968). Inadequacies in this theory at low electron densities, similar to those encountered in positron lifetime calculations (Sjölander and Stott 1970) were patched up with a plausible interpolation procedure for which the justification has now been established by Bhattacharyya and Singwi (1972 b). The thus confirmed correlation energies decreased smoothly from -1 Ryd at $r_s \sim 1$ A.U. to a low density ($r_s > 4$ A.U.) limiting value of -0.5 Ryd appropriate to a bound positronium system. No account was taken of lattice or core annihilation effects but analogy with the model discussed in § 2.4.3 suggests a simple approximate way in which the core polarization might be included.

Zero-point or ground-state energies due to positron–ion interaction were computed in the Wigner–Seitz approximation and were, in the main, in good agreement with earlier calculations by Hodges (1970). Calculations of the surface dipole contribution, which, for positrons, acts in the same sense as the zero-point energy term, involved a variety of considerations for which this article has not supplied the groundwork. The reader is referred to the original account for details.

The accumulated results of the Hodges and Stott calculations are shown in table 11. Negative positron work functions

$$\phi_{\mathrm{p}} = -(E_{\mathrm{corr}} + E_0 + D)$$

are obtained for copper and gold, but only for gold do the authors claim confidence for the sign of their result. Hodges and Stott also pose the interesting question of whether it is really necessary to strip the positron of its polarization cloud in order to remove it from the metal. A positronium work function is defined in terms of the positron work function ϕ_{p}, the electron work function ϕ_{e}, and the positronium binding energy, by the plausible relation

$$\phi_{\mathrm{ps}} = \phi_{\mathrm{p}} + \phi_{\mathrm{e}} - 0\cdot 5 \ \mathrm{Ryd.}$$

The deduced, and sometimes large, negative values of ϕ_{ps} obtained (table 11) suggest again the possibility of positronium formation in voided materials. It is perhaps unfortunate that molybdenum was not a subject of these studies.

It would not be rational to conclude this section without considering the pertinent question of to what extent positron studies, ostensibly in bulk material, are, or could be affected by surface effects. Schulte *et al.* (1972) have reported a particular sensitivity of lineshape measurements to such effects when low energy positrons from ^{58}Co or ^{22}Na sources are employed. Similar effects in lifetime studies (Kugel *et al.* 1966) were noted in § 1.3.1. It is this writer's experience that surface contamination has to be severe before it can be seen in angular correlation studies employing a separated source-sample geometry (fig. 4) and we would assert, therein lies a clue.

The backscattering of energetic positrons off any sample surface is (Bertolaccini and Zappa 1967, MacKenzie *et al.* 1973) like the absorption (Seliger 1955, Takhar 1967, Rupaal and Patrick 1972), a function of atomic number. The transfer of positrons between materials of different atomic is most difficult when passing from low Z to high Z materials (Bertolaccini and Zappa 1967). The sandwich arrangement, usually employed in lifetime and annihilation lineshape studies, may encourage multiple reflections between the sample layers with ultimate annihilation in source containing foils, or sample surface layers. Thus, with such

Table 11. The positron zero point energy E_0, the positron–electron correlation energy E_{corr}, and the dipole layer contribution D. The last two columns list the estimated values for the positron work function, ϕ_p and the positronium work function, ϕ_{ps} (Hodges and Stott 1973)

	r_s(a.u.)	E_0(Ryd)	E_{corr}(Ryd)	D(Ryd)	$\phi_p = -E_{corr}$ $-E_1 - D$ (Ryd)	$\phi_{ps} = \phi_e + \phi_p$ -0.5 (Ryd)
Li	3·26	0·13	−0·54	0·09	+0·32	+0·05
Na	3·93	0·13	−0·52	0·05	+0·34	+0·04
K	4·86	0·10	−0·51	0·02	+0·38	+0·06
Rb	5·20	0·09	−0·51	0·01	+0·40	+0·06
Cs	5·63	0·08	−0·51	−0·00	+0·42	+0·08
Mg	2·65	0·23	−0·58	0·15	+0·20	−0·03
Zn	2·30	0·34	−0·62	0·21	+0·07	−0·11
Cd	2·59	0·32	−0·59	0·14	+0·13	−0·06
Hg	2·66	0·31	−0·58	0·06	+0·21	+0·04
Al	2·07	0·35	−0·65	0·25	+0·05	−0·14
Ga	2·19	0·33	−0·63	0·15	+0·15	−0·02
In	2·41	0·31	−0·60	0·10	+0·19	−0·02
Tl	2·48	0·30	−0·59	0·03	+0·26	+0·04
Sn	2·21	0·34	−0·63	0·09	+0·20	+0·02
Pb	2·30	0·32	−0·62	0·02	+0·28	+0·07
Cu	2·67	0·31	−0·58	0·34	−0·07	−0·23
Ag	3·01	0·31	−0·55	0·22	+0·02	−0·19
Au	3·01	0·34	−0·55	0·33	−0·12	−0·24

geometries, sources ([68]Ge, [44]Sc) emitting more energetic positrons, may be preferred (Schulte et al. 1972).

Geometrical problems have also been suggested (Campbell et al. 1972) for the origin of the anomalous effect of temperature on angular correlation reported by some workers. When many metallic single crystal samples are cooled to $\sim 100°$K a decrease in the area under the two photon angular distribution is observed (Dekhtyar and Mikhalenkov 1962, Faraci Foti, Quercia and Turrisi 1969, 1970). No such effect is observed for polycrystalline samples.

Dekhtyar (1969) proposed that positron channelling (Uggerhøj and Andersen 1968, Armagan OK 1970), inhibited at room temperature by lattice vibrations, allowed for positron escape from cooled single crystals. Subsequent angular correlation experiments (Dekhtyar et al. 1971) seemed to support this hypothesis but ingenious lifetime studies by Faraci, Foti and Turrisi (1970) were unable to detect the escape of positrons from the crystals.

Faraci, Foti, Quercia and Turrisi (1970) have suggested an alternative hypothesis in which positron channelling results in an enhanced annihilation of positrons of high energy and thus momenta, lying outside the normally scanned angular range. Again, more recent experiments cast doubt on this hypothesis. Campbell *et al.* (1972) find, in Doppler-broadening studies, no evidence of an increase in large energy annihilations when the appropriate single crystals are cooled.

It must be admitted that not everyone is able to observe the effects in angular correlation (Brandeis University Positron Group 1971). Nevertheless it is not unknown for preliminary reports of unexpected phenomena to become the subject of controversy. A common, relatively uncomplicated delocalized positron state served to explain the experimental results for metals for more than a decade. Our comparatively recent discovery of more complex positron behaviour has already brought in its wake a variety of applications. The era of real discovery is not yet ended and fundamental investigations of positron states must continue to play an important part in positron studies in matter.

Acknowledgments

I would like to thank my family and various colleagues for their tolerance and good humour during the preparation of this article. I am particularly grateful to Professor W. H. Young and Dr. V. H. C. Crisp who kindly read through parts of the manuscript and supplied many helpful comments and criticisms. Thanks are also due to the many authors and publishers who permitted the reproduction of figures and tables : the sources are indicated in the text.

References

ADAMENKO, A. A., and SHALAEV, A. M., 1971, *Phys. Stat. Sol.* B, **43,** K147.

AKAHANE, T., MORINAGA, H., SUEOKA, O., and FUJIWARA, K., 1971, *Proc. Second Int. Conf. on Positron Annihilation*, Kingston, Ontario, September, p. 1.143 (unpublished).

AKHIEZER, A. I., and BERESTETSKII, V. B., 1965, *Quantum Electrodynamics* (New York Interscience Publishers).

AREFIEV, K. P., and VOROBIEV, S. A., 1972, *Phys. Lett.* A., **39,** 381.

ARIAS-LIMONTA, J. A., and VARLASHKIN, P. G., 1970 a, *Phys. Rev.* B, **1,** 142 ; 1970 b, *J. chem. Phys.*, **52,** 581.

ARMAGAN OK, H. H., 1970, *Z. Phys.*, **240,** 314.

ARPONEN, J., 1970, *J. Phys.* C, **3,** 107 ; 1971, *Proc. Second Int. Conf. on Positron Annihilation*, Kingston, Ontario, September, p. 1.176 (unpublished).

ARPONEN, J., HAUTOJARVI, P., NIEMINEN, R., and PAJANNE, E., 1973, *Solid St. Commun.*, **12,** 143.

ARPONEN, J., and JAUHO, P., 1968, *Phys. Rev.*, **167,** 239.

BADOUX, F., and HEINRICH, R., 1970, *Helv. phys. Acta*, **43,** 473.

BALLENTINE, L. E., 1966, *Can. J. Phys.*, **44,** 2533.

BAY, Z., 1960, *Phys. Rev.*, **72,** 419.

BELL, R. E., 1966, *Nucl. Instrum. Methods*, **55,** 1.

BELL, R. E., and GRAHAM, R. L., 1953, *Phys. Rev.*, **90,** 644.

BERESTETSKII, V. B., LIFSHITZ, E. M., and PITAEVSKII, L. P., 1971, *Relativistic Quantum Theory* (Oxford: Pergamon).

BERGERSEN, B., 1964, Ph.D. Thesis, Brandeis University (unpublished) ; 1969, *Phys. Rev.*, **181,** 499.

BERGERSEN, B., and MCMULLEN, T., 1971, *Solid St. Commun.*, **9,** 1865.

BERGERSEN, B., and PAJANNE, E., 1969, *Phys. Rev.*, **186,** 375 ; 1971, *Ibid.*, **133,** 1588.

BERGERSEN, B., and STOTT, M. J., 1969, *Solid St. Commun.*, **7,** 1203.

BERGERSEN, B., and TERRELL, J. H., 1968, *Soft X-ray Band Spectra and the Electronic Structure of Metals and Materials*, edited by D. J. Fabian (New York : Academic Press), p. 351.

BERKO, S., 1962, *Phys. Rev.*, **127,** 2166 ; 1967, *Positron Annihilation* (New York : Academic Press), p. 61.

BERKO, S., CUSHNER, S., and ERSKINE, J. C., 1968, *Phys. Lett.* A, **27,** 668.

BERKO, S., and ERKSINE, J. C., 1967, *Phys. Rev. Lett.*, **19,** 307.

BERKO, S., KELLY, R. E., and PLASKETT, J. S., 1957, *Phys. Rev.*, **106,** 824.

BERKO, S., and MILLS, A. P., 1971, *J. Phys.*, *Paris*, **32,** C1–287.

BERKO, S., and PLASKETT, J. S., 1958, *Phys. Rev.*, **112,** 1877.

BERTOLACCINI, M., BISI, A., GAMBARINI, G., and ZAPPA, L., 1971, *J. Phys.* C, **4,** 734.

BERTOLACCINI, M., and DUPASQUIER, A., 1970, *Phys. Rev.* B, **1,** 2896.

BERTOLACCINI, M., and ZAPPA, L., 1967, *Nuovo Cim.* B, **50,** 256.

BETHE, H., and GOLSTONE, J., 1957, *Proc. R. Soc.* A, **238,** 551.

BHATTACHARYYA, F., and SINGWI, K. S., 1972 a, *Phys. Rev. Lett.*, **29,** 22 ; 1972 b, *Physics Lett.* A, **41,** 457.

BHIDE, M. G., and MAJUMDAR, C. K., 1969, *J. Phys.* B, **2,** 966.

BISI, A., BUSSOLATI, C., COVA, S., and ZAPPA, L., 1966, *Phys. Rev.*, **141,** 348.

BISI, A., DUPASQUIER, A., and ZAPPA, L., 1971 a, *Proc. Second Int. Conf. on Positron Annihilation*, Kingston, Ontario, September, p. 2.21 (unpublished) ; 1971 b, *J. Phys.* C, **4,** L33.

BISI, A., FIORENTINI, A., and ZAPPA, L., 1963, *Phys. Rev.*, **131,** 1023 ; 1964, *Ibid.*, **134,** A 328.

BISI, A., GAMBARINI, G., and ZAPPA, L., 1970, *Nuovo Cim.* B, **67,** 75 ; 1971, *Phys. Lett.* A, **35,** 193.

BRANDEIS UNIVERSITY POSITRON GROUP, 1971, *Proc. Second Int. Conf. on Positron Annihilation*, Kingston, Ontario, September, p. 1.175 (unpublished).

BRANDT, W., 1967, *Positron Annihilation* (New York : Academic Press), p. 155.

BRANDT, W., BERKO, S., and WALKER, W. W., 1960, *Phys. Rev.*, **120,** 1289.

BRANDT, W., COUSSOT, G., and PAULIN, R., 1969, *Phys. Rev. Lett.*, **23,** 522.

BRANDT, W., EDER, L., and LUNDQUIST, S., 1966, *Phys. Rev.*, **142,** 165.

BRANDT, W., and FAHS, J. H., 1970, *Phys. Rev.*, B **2,** 1425.

BRANDT, W., and FEIBUS, H., 1968, *Phys. Rev.*, **174,** 454 ; 1969, *Ibid.*, **184,** 277.

BRANDT, W., and PAULIN, R., 1968, *Phys. Rev. Lett.*, **21,** 193 ; 1972, *Phys. Rev.* B, **5,** 2430.

BRANDT, W., and REINHEIMER, J., 1971, *Phys. Lett.* A, **35,** 109.

BRANDT, W., and SPIRN, I., 1966, *Phys. Rev.*, **142,** 231.

BRANDT, W., and WAUNG, H. F., 1968, *Phys. Lett.* A, **27,** 700 ; 1971, *Phys. Rev.*, **133,** 3432.

BRANDT, W., WAUNG, H. F., and LEVY, P. W., 1968, *Proc. Int. Symposium on Colour Centres in Alkali Halides*, Rome, p. 48 ; 1971, *Phys. Rev. Lett.*, **26,** 496.

BRISCOE, C. V., CHOI, S. I., and STEWART, A. T., 1968, *Phys. Rev. Lett.*, **20,** 493.

BRISCOE, C. V., and STEWART, A. T., 1967, *Positron Annihilation* (New York : Academic Press), p. 377.

BURTON, J. J., and JURA, G., 1968, *Phys. Rev.*, **171,** 699.

BUSSOLATI, C., DUPASQUIER, A., and ZAPPA, L., 1967, *Nuovo Cim.*, B **52,** 529.

CALLAWAY, J., 1959, *Phys. Rev.*, **116,** 1140.

CAMPBELL, J. L., SCHULTE, C. W., and MACKENZIE, I. K., 1972, *Phys. Lett.* A, **38,** 377.

CANGAS, A., JELENSKA-PEINKOWSKA, H., SWIATKOWSKI, W., and WESOLOWSKI, J., 1967, *Acta phys. pol.*, **32**, 719.

CARBOTTE, J. P., 1966, *Phys. Rev.*, **144**, 309 ; 1967, *Ibid.*, **155**, 197.

CARBOTTE, J. P., and ARORA, H. L., 1967, *Can. J. Phys.*, **45**, 387.

CARBOTTE, J. P., and KAHANA, S., 1965, *Phys. Rev.*, **139**, A213.

CARBOTTE, J. P., and SALVADORI, A., 1967, *Phys. Rev.*, **162**, 290.

CHANG LEE, 1958, *Soviet Phys.*, *JETP*, **6**, 281.

CHIBA, T., NUGOCHI, M., MITSUHASHI, T., TANAKA, T., and TSUDA, N., 1971, *J. Phys. Soc. Japan*, **31**, 1288.

CHOLLET, L. F., and TEMPLETON, I. M., 1968, *Phys. Rev.*, **170**, 656.

CHOUINARD, M. P., GUSTAFSON, D. R., and HECKMAN, R. C., 1969, *J. chem. Phys.*, **51**, 3554.

CHUANG, S. Y., and HOGG, B. G., 1967, *Can. J. Phys.*, **45**, 3895.

CHUANG, S. Y., HOLT, W. H., and HOGG, B. G., 1968, *Can. J. Phys.*, **46**, 2309

CHUANG, S. Y., and TAO, S. J., 1970, *Phys. Lett.* A, **33**, 56.

CHUANG, S. Y., TAO, S. J., and WILKENFIELD, J. M., 1971, *Proc. Second Int. Conf. on Positron Annihilation*, Kingston, Ontario, September, p. 5.68 (unpublished) ; 1972, *J. appl. Phys.*, **43**, 737.

COLE, G. D., and WALKER, W. W., 1965, *J. chem. Phys.*, **42**, 1692.

COLOMBINO, P., FISCELLA, B., and TROSSI, L. 1963, *Nuovo Cim.*, **27**, 589 ; 1964, *Ibid.*, **31**, 950 ; 1965, *Ibid.*, **38**, 707.

CONNORS, D. C., and BOWLER, J., 1973, *Physics Lett.* A, **43**, 395.

CONNORS, D. C., CRISP, V. H. C., and WEST, R. N., 1970, *Physics Lett.* A, **33**, 180 ; 1971, *J. Phys.* F, **1**, 355.

CONNORS, D. C., and WEST, R. N., 1969, *Physics Lett.* A, **30**, 24.

COOPER, A. M., DEBLONDE, G., and HOGG, B. G., 1969, *Phys. Lett.* A, **29**, 275.

COOPER, M., 1971, *Adv. Phys.*, **20**, 452.

COOPER, A. M., LAIDLAW, G. J., and HOGG, B. G., 1967, *J. chem. Phys.*, **46**, 2441.

COSTELLO, D. G., GRACE, D. E., HERRING, D. F., and McGOWAN, J. W., 1972, *Phys. Rev.* B, **5**, 1433.

COTTERILL, R. M. J., MACKENZIE, I. K., SMEDSKJAER, L., TRUMPY, G., and TRÄFF, J. H. O., 1972, *Nature, Lond.*, **239**, 99.

COTTERILL, R. M. J., PETERSEN, K., TRUMPY, G., and TRÄFF, J., 1972, *J. Phys.* F, **2**, 459.

COUSSOT, G., 1969, *Phys. Lett.* A, **30**, 138.

COVA, S., and ZAPPA, L., 1968, *J. Phys.* B, **1**, 795.

CRACKNELL, A. P., 1969, *Adv. Phys.*, **18**, 703.

CRISP, V. H. C., LOCK, D. G., and WEST, R. N., 1973, *J. Phys.* F (in the press).

CROWELL, J., ANDERSON, V. E., and RITCHIE, R. H., 1966, *Phys. Rev.*, **150**, 243.

CUSHNER, S., ERSKINE, J. C., and BERKO, S., 1970, *Phys. Rev.* B, **1**, 2852.

DANIEL, E., and VOSKO, S. H., 1960, *Phys. Rev.*, **120**, 2041.

DAVE, N. K., McKEE, B. T. A., STEWART, A. T., STOTT, M. J., and TRIFTHAÜSER, W., 1971, *Proc. Second Int. Conf. on Positron Annihilation*, Kingston, Ontario, September, p. 4.49 (unpublished).

DE BENEDETTI, S., COWAN, C. E., KONNEKER, W. R., and PRIMAKOFF, H., 1950, *Phys. Rev.*, **77**, 205.

DE BLONDE, G., CHUANG, S. Y., HOGG, B. G., KERR, D. P., and MILLER, C. M., 1972, *Can. J. Phys.*, **50**, 1619.

DEKHTYAR, I. YA., 1969, *Physics Lett.* A, **30**, 462 ; 1970, *Ibid.*, **31**, 546 ; 1971, *Phys. Stat. Sol.* V, **48**, K47.

DEKHTYAR, I. YA., and CIZEK, A., 1971, *Physics Lett.* A, **34**, 345.

DEKHTYAR, I. YA., LEVINA, D. A., and MIKHALENKOV, V. S., 1964, *Soviet Phys. Dokl.*, **9**, 492.

DEKHTYAR, I. YA., and MIKHALENKOV, V. S., 1961, *Soviet Phys. Dokl.*, **6**, 31; 1962, *Ibid.*, **6**, 917 ; 1971, *Phys. Stat. Sol.* B, **47**, K121.

DEKHTYAR, I. YA., MIKHAELENKOV, V. S., and CIZEK, A., 1971, *Phys. Stat. Sol.* B, **47**, K117.

DEKHTYAR, I. YA., and SHEVCHENKO, V. S., 1972, *Phys. Stat. Sol.* B, **49**, K11.

DEKHTYAR, I. YA., SILANTEV, V. I., and ADONKIN, V. T., 1972, *Phys. Stat. Sol.* A, **11**, K153.

DE ZAFRA, R. L., and JOYNER, W. T., 1958, *Phys. Rev.*, **112**, 19.

DIRAC, P. A. M., 1930, *Proc. Camb. phil. Soc. math. phys. Sci.*, **26**, 361.

DONAGHY, J. J., and STEWART, A. T., 1967 a, *Phys. Rev.*, **164**, 391, 1967 b, *Ibid.*, **164**, 396.

DONOVAN, B., and MARCH, N. H., 1958, *Phys. Rev.*, **110**, 582.

DUBOIS, D. F., 1959, *Ann. Phys.*, **7**, 174 ; *Ibid.*, **8**, 24.

DUPASQUIER, A., 1970, *Nuovo. Cim. Lett.*, **4**, 13.

EHRENREICH, H., and HODGES, L., 1968, *Methods of Computational Physics*, Vol. 8 (New York : Academic Press).

ELDRUP, M., MOGENSEN, O. E., PETERSEN, K., and TRUMPY, G., 1971k *Proc. Second Int. Conf. on Positron Annihilation*, Kingston, Ontario, September, p. 3.63 (unpublished).

ERSKINE, J. C., and McGERVEY, J. D., 1966, *Phys. Rev.*, **151**, 615.

FABIAN, D. J., 1968, *Soft X-ray Band Spectra and the Electronic Structure of Metals and Materials* (New York : Academic Press), Part 2.

FABRI, G., GERMAGNOLI, E., IACI, G., QUERCIA, I. E., and TURRISI, E., 1967, *Positron Annihilation*, edited by Stewart and Roellig (New York : Academic Press), p. 357.

FABRI, G., GERMAGNOLI, E., QUERCIA, I. F., and TURRISI, E., 1963, *Nuovo Cim.*, **30**, 21.

FARACI, G., FOTI, G., QUERCIA, I. F., and TURRISI, E., 1970, *Phys. Rev.* B, **2**, 2335.

FARACI, G., FOTI, G., and TURRISI, E., 1970, *Physics Lett.* A, **31**, 427.

FARACI, G., QUERCIA, I. F., SPADONI, M., and TURRISI, E., 1969, *Nuovo Cim.* B, **60**, 228.

FERRANTE, G., 1968, *Phys. Rev.*, **170**, 76.

FERRANTE, G., and GERACITANO, R., 1970, *Nuovo Cim. Lett.*, **3**, 48.

FERRELL, R. A., 1956, *Rev. mod. Phys.*, **28**, 308 ; 1957, *Phys. Rev.*, **108**, 167; 1958, *Ibid.*, **110**, 1355.

FIESCHI, R., GAINOTTI, A., GHEZZI, C., and MANFREDI, M., 1968, *Phys. Rev.*, **175**, 383.

FOLDY, L. L., and WOUTHUYSEN, S. A., 1950, *Phys. Rev.*, **78**, 29.

FUJIWARA, K., 1970, *J. Phys. Soc. Japan*, **29**, 1479.

FUJIWARA, K., and SUEOKA, O., 1966, *J. phys. Soc. Japan*, **29**, 1479.

FUJIWARA, K., SUEOKA, O., and IMURA, T., 1968, *J. Phys. Soc. Japan*, **24**, 467.

GAINOTTI, A., GERMOGNOLI, E., SCHIANCHI, G., and ZECCHINA, L., 1964, *Physics Lett.*, **13**, 9.

GAINOTTI, A., and GHEZZI, C., 1970, *Phys. Rev. Lett.*, **24**, 349.

GEDCKE, D. A., and McDONALD, W. J., 1968, *Nucl. Instrum. Methods*, **58**, 253.

GEDCKE, D. A., and WILLIAMS, C. W., 1968, Ortec Information Sheet—High Resolution Time Spectroscopy. 1. Scintillation Detectors.

GEMPEL, R. F., GUSTAFSON, D. R., and WILLENBERG, J. D., 1972, *Phys. Rev.* B, **5**, 2082.

GERMAGNOLI, E., POLETTI, G., and RANDONE, G., 1966, *Phys. Rev.*, **141**, 119.

GOLDANSKII, V. I., 1967, *Positron Annihilation* (New York : Academic Press), pp. 183–255 ; 1968, *Atom. Energy, Rev.*, **6**, 3.

GOLDANSKII, V. I., and PROKOP'EV, E. P., 1965, *Soviet Phys.—Solid St.*, **6**, 2641 ; 1966, *Ibid.*, **6**, **8**, 409.

GOULD, A. G., WEST, R. N., and HOGG, B. G., 1972, *Can. J. Phys.*, **50**, 2294.

GRAY, P. R., COOK, C. F., and STURM, G. P., 1968, *J. chem. Phys.*, **48**, 1145.

GREEN, J. H., and LEE, J., 1964, *Positronium Chemistry*, (New York : Academic Press).

GREEN, R., BOS, W. G., and HUANG, W. F., 1971, *Phys. Rev.* B, **3**, 64.

GREENBERGER, A., MILLS, A. P., THOMPSON, A., and BERKO, S., 1970, *Physics Lett.* A, **32**, 72.

GROSSKREUTZ, J. C., and MILLET, W. E., 1969, *Physics Lett.* A, **28**, 621.

GUSTAFSON, D. R., and MACKINTOSH, A. R., 1963, *Physics Lett.*, **5**, 234 ; 1964, *J. Phys. Chem. Solids*, **25**, 389.

GUSTAFSON, D. R., MACKINTOSH, A. R., and ZAFFARANO, D. J., 1963, *Phys. Rev.*, **130**, 1455.

GUSTAFSON, D. R., McNUTT, J. D., and ROELLIG, L. O., 1969, *Phys. Rev.*, **183**, 435.

HALPERN, O., 1954, *Phys. Rev.*, **94**, 904.

HALSE, M. R., 1969, *Phil. Trans. R. Soc.*, **268**, 507.

HAMANN, D. R., 1966, *Phys. Rev.*, **146**, 277.

HARTREE, D. R., 1952, *Numerical Analysis* (Oxford : Clarendon Press).

HATANO, A., KANAZAWA, H., and MIZUINO, Y., 1965, *Prog. theor. Phys.*, **34**, 874.

HAUTOJÄRVI, P., 1972, *Solid St. Commun.*, **11**, 1049.

HAUTOJÄRVI, P., and JAUHO, P., 1967, *Physics Lett.* A, **25**, 729 ; 1971, *Proc. Second Int. Conf. on Positron Annihilation*, Kingston, Ontario, September, p. 4.15 (unpublished).

HAUTOJÄRVI, P., TAMMINEN, A., and JAUHO, P., 1970, *Phys. Rev. Lett.*, **24**, 459.

HEDE, B. B. J., and CARBOTTE, J. P., 1962, *J. Phys. Chem. Solids*, **33**, 727.

HELD, A., and KAHANA, S., 1964, *Can. J. Phys.*, **42**, 1968.

HERLACH, D., and HEINRICH, F., 1970 a, *Physics Lett.* A, **31**, 47 ; 1970 b, *Helv. Phys. Acta*, **43**, 489.

HERMAN, F., and SKILLMAN, S., 1963, *Atomic Structure Calculations* (Prentice-Hall).

HERNANDEZ, J. P., and CHOI, S., 1969, *Phys. Rev.*, **188**, 340.

HODGES, C. H., 1970, *Phys. Rev. Lett.*, **25**, 284.

HODGES, C. H., McKEE, B. T. A., TRIFTHÄUSER, W., and STEWART, A. T., 1972, *Can. J. Phys.*, **50**, 103.

HODGES, C. H., and STOTT, M. J., 1973, *Phys. Rev.* B, **7**, 73.

HOGG, B. G., and LAIDLAW, G. M., 1968,, *Atom. Energy Rev.*, **6**, 149.

HOHENEMSER, C., WEINGART, J. M., and BERKO, S., 1968, *Physics Lett.*, A **28**, 41.

HOLT, W. H., CHUANG, S. Y. COOPER, A. M., and HOGG, B. G., 1968, *J. chem. Phys.*, **49**, 5147.

HOLT, W. H., ROSE, M. F., CHUANG, S. Y., and TAO, S. J., 1970, *Physics Lett*, A, **32**, 422.

HOTZ, H. P., MATHIESEN, J. M., and HURLEY, J. P., 1968, *Phys. Rev.*, **170**, 351.

HOWELLS, M. R., and OSMON, P. E., 1972, *J. Phys.* F, **2**, 277.

HSU, F. H., MALLARD, W. C., and HADLEY, J. H., 1971, *Proc. Second Int. Conf. on Positron Annihilation*, Kingston, Ontario, September, p. 2.12 (unpublished).

HSU, F. H. H., and WU, C. S., 1967, *Phys. Rev. Lett.*, **18**, 889.

HUBBARD, J., 1957, *Proc. R. Soc.* A, **243**, 336.

HUME-ROTHERY, W., and RAYNOR, G. V., 1938, *Phil. Mag.*, **26**, 129.

HYODO, T., SUEOKA, O., and FUJIWARA, K., 1971, *J. phys. Soc. Japan*, **31**, 563.

ITOH, F., KUROHA, M., KAI, K., and TAKEUCHI, S., 1972, *J. Phys. Soc. Japan*, **33**, 567.

JONES, H., 1960, *Theory of Brillouin Zones and Electronic States in Crystals* (Amsterdam : North Holland).

KAHANA, S., 1963, *Phys. Rev.*, **129**, 1622 ; 1967, *Positron Annihilation* (New York : Academic Press).

KANAZAWA, H., HOTSUKI, Y., and YANAGAWA, S., 1965, *Prog. theor. Phys.*, **33**, 1010.

KEETON, S. C., and LOUCKS, T. L., 1968, *Phys. Rev.*, **168**, 672.

KELLY, T. M., 1971, *Proc. Second Int. Conf. on Positron Annihilation*, Kingston, Ontario, September, p. 2.43 (unpublished).

KERR, D. P., CHUANG, S. Y., and HOGG, B. G., 1965, *Molec. Phys.*, **10**, 13.

KERR, D. P., and HOGG, B. G., 1962, *J. chem. Phys.*, **36**, 2109.

KIM, S. M., and BUYERS, W. J. L., 1972, *Can. J. Phys.*, **50**, 1777.

KIM, S. M., STEWART, A. T., and CARBOTTE, J. P., 1967, *Phys. Rev. Lett.*, **18**, 385.

KIRKEGAARD, P., and ELDRUP, M., 1971, Danish Atomic Energy Commission, Risø, Denmark, Report No. Risø–M–1400.

KIRKEGAARD, P., and ELDRUP, M., 1972, *Computer Phys. Commun.*, **3**, 240.

KNAPTON, D., and McKEE, B. T. A., 1971, *Proc. Second Int. Conf. on Positron Annihilation*, Kingston, Onatrio, September, p. 470 (unpublished).

KUBICA, P., McKEE, B. T. A., STEWART, A. T., and STOTT, M. J., 1971, *Proc. Second Int. Conf. on Positron Annihilation*, Kingston, Ontario, September, p. 1.153 (unpublished).

KUGEL, H. W., FUNK, E. G., and MIHELICH, J. W., 1966, *Physics Lett.*, **20**, 364.

KURIBAYASHI, K., TANIGAWA, S., NANAO, S., and DOYAMA, M., 1972, *Physics Lett. A*, **40**, 27.

KUSMISS, J. H., 1965, Ph.D. Thesis, University of North Carolina (unpublished).

KUSMISS, J. H., ESSELTINE, C. D., SNEAD, C. L., and GOLAND, A. N., 1970, *Physics Lett. A*, **32**, 175.

KUSMISS, J. H., and STEWART, A. T., 1967 a, *Adv. Phys.*, **16**, 63 ; 1967 b, *Positron Annihilation* (New York : Academic Press), p. 341.

KUSMISS, J. H., and SWANSON, J. W., 1968, *Physics Lett. A*, **27**, 517.

LAGU, R. G., KULKARNI, V. G., THOSAR, B. V., and CHANDRA, G., 1969, *Proc. Indian Acad. Sci.*, **69**, 48.

LANDES, H. S., BERKO, S., and ZUCHELLI, A. J., 1956, *Phys. Rev.*, **102**, 724.

LANG, G., and DE BENEDETTI, S., 1957, *Phys. Rev.*, **108**, 257.

LANZCOS, C., 1957, *Applied Analysis* (London : Pitman), p. 272.

LEE, J., and CELITANS, G. J., 1965, *J. chem. Phys.*, **42**, 437 ; 1966, *Ibid.*, **44**, 2506.

LEE-WHITING, G. E., 1955, *Phys. Rev.*, **91**, 1557.

LICHTENBERGER, P. C., STEVENS, J. R., and NEWTON, T. D., 1971, *Proc. Second Int. Conf. on Positron Annihilation*, Kingston, Ontario, September, p. 4.104 (unpublished).

LIDIARD, A. B., 1957, *Handbuch der Physik*, Vol. 20, edited by S. Flügge (Berlin : Springer), .p 298.

LIU, D. C., and ROBERTS, W. K., 1963, *Phys. Rev.*, **132**, 1633.

LOCK, D. G., CRISP, V. H. C., and WEST, R. N., 1973, *J. Phys. F*, **3**, 561.

LOUCKS, T. L., 1966, *Phys. Rev.*, **144**, 504.

McGERVEY, J. D., 1967, *Positron Annihilation* (New York : Academic Press), p. 143.

McGERVEY, J. D., and WALTERS, V. F., 1970, *Phys. Rev. B*, **2**, 2421.

McKEE, B. T. A., JOST, A. G. D., and MACKENZIE, I. K., 1972, *Can. J. Phys.*, **50**, 415.

McKEE, B. T. A., LANGSTROTH, G. F. O., and MACKENZIE, I. K., 1967, *Positron Annihilation* (New York : Academic Press), p. 281.

McKEE, B. T. A., TRIFTSHÄUSER, W., and STEWART, A. T., 1972, *Phys. Rev. Lett.*, **28**, 358.

MACKENZIE, I. K., CRAIG, T. W., and McKEE, B. T. A., 1971, *Physics Lett.*, A, **36**, 227.

MACKENZIE, I. K., EADY, J. A., and GINGERICH, R. R., 1970, *Physics Lett*, A, **33**, 279.

MACKENZIE, I. K., GINGERICH, R. R., and KIM, S. M., 1971, *Proc. Second Int. Conf. on Positron Annihilation*, Kingston, Ontario, September, p. 4.59 (unpublished).

MACKENZIE, I. K., KHOO, T. L., McDONALD, A. B., and McKEE, B. T. A., 1967, *Phys. Rev. Lett.*, **19,** 946.

MACKENZIE, I. K., LANGSTROTH, G. F. O., McKEE, B. T. A., and WHITE, C. G., 1964, *Can. J. Phys.*, **42,** 1837.

MACKENZIE, I. K., LeBLANC, R., and McKEE, B. T. A., 1971, *Phys. Rev. Lett.*, **27,** 580.

MACKENZIE, I. K., SCHULTE, C. W., JACKMAN, T., and CAMPBELL, J. L., 1973, *Phys. Rev. A*, **7,** 135.

MAJUMDAR, C. K., 1965 a, *Phys. Rev. A*, **140,** 227 ; 1965 b, *Ibid.*, **140,** 235 ; 1966, *Ibid.*, **149,** 406 ; 1971, *Ibid.*, B, **4,** 2111.

MAJUMDAR, C. K., and RAJAGOPAL, A. K., 1970, *Prog. theor. Phys.*, **44,** 26.

MALLARD, W. C., and HSU, F. H., 1972, *Physics Lett. A*, **38,** 164.

MARTIN, J. W., and PAETSCH, R., 1972, *J. Phys. F*, **2,** 997.

MATTUCK, R. D., 1967, *A Guide to Feynmann Diagrams in the Many-Body Problem* (New York : McGraw-Hill).

MELNGAILIS, J., 1970, *Phys. Rev. B*, **2,** 563.

MELNGAILIS, J., and DE BENEDETTI, S., 1966, *Phys. Rev.*, **145,** 400.

MICAH, E. T., and YOUNG, W. H., 1970, *Physics Lett. A*, **33,** 391.

MIHALISIN, T. W., and PARKS, R. D., 1969, *Solid St. Commun.*, **7,** 32.

MIJNARENDS, P. E., 1967, *Phys. Rev.*, **160,** 512 ; 1969, *Ibid.*, **178,** 622 ; 1973 a, *Physica*, **63,** 235 ; 1973 b, *Ibid.*, **63,** 248.

MIKESKA, H., 1967, *Physics Lett. A*, **24,** 402 ; 1970, *Z. Phys.*, **232,** 159.

MILLET, W. E., DIETERMAN, L. H., and THOMPSON, J. C., 1967, *Positron Annihilation* (New York : Academic Press), p. 317.

MILLS, A. P., 1968, *Physics Lett. A*, **26,** 286.

MOGENSEN, O. E., 1970, *Nucl. Instrum. Methods*, **84,** 293.

MOGENSEN, O. E., KVAJIC, G., ELDRUP, M., and MILOSEVIC-KVAJIC, M., 1971, *Phys. Rev. B*, **4,** 71.

MOGENSEN, O. E., PETERSEN, K., COTTERILL, R. M. J., and HUDSON, B., 1972, *Nature, Lond.*, **239,** 98.

MOGENSEN, O. E., and TRUMPY, G., 1969, *Phys. Rev.*, **188,** 639.

MOTT, N. F., and JONES, H., 1936, *The Theory of the Properties of Metals and Alloys* (New York : Dover Publications).

MOTT, N. F., and ZINAMON, Z., 1970, *Rep. Prog. Phys.*, **33,** 881.

MURRAY, B. W., and McGERVEY, J. D., 1970, *Phys. Rev. Lett.*, **24,** 9.

NAKAI, H., HASEQAWA, M. and WATABE, M., 1973, *The Properties of Liquid Metals* (London : Taylor & Francis Ltd.), p. 209.

NANAO, S., KURIBAYASHI, K., TANIGAWA, S., and DOYAMA, M., 1972, *Physics Lett. A*, **38,** 489.

NEAMTAN, S. M., DAREWYCH, G., and OCZKOWSKI, G., 1962, *Phys. Rev.*, **126,** 193.

NEAMTAN, S. M., and VERRALL, R. I., 1964, *Phys. Rev. A*, **134,** 1254.

ORE, A., 1949, Univ. i Bergen Arbok, Naturvitenskap. Rekke, No. 9.

PACIGA, J. J., and WILLIAMS, D. Ll., 1971, *Can. J. Phys.*, **49,** 3227.

PAGE, L. A., and HEINBERG, M., 1956, *Phys. Rev.*, **102,** 1545.

PAUL, D. A. L., 1958, *Can. J. Phys.*, **36,** 640.

PAULIN, R., and AMBROSINO, G., 1967, *Positron Annihilation*, (New York : Academic Press), p. 345.

PENDYALA, S., ORTH, P. H. R., McGOWAN, J. W., and ZITEZWITZ, P. W., 1971, *Proc. Second Int. Conf. on Positron Annihilation*, Kingston, Ontario, September, p. 4.140 (unpublished).

PICK, H., 1972, *Optical Properties of Solids*, edited by F. Abeles (Amsterdam : North-Holland), p. 653.

PINES, D., 1962, *The Many Body Problem* (New York : Benjamin) ; 1963, *Elementary Excitations in Solids* (New York : W. A. Benjamin), p. 34.

PIRENNE, J., 1947, *Archs. Sci. phys. nat.*, **29**, 293.

PRANGE, R., and KLEIN, A., 1958, *Phys. Rev.*. **112**, 1008.

PROKOP'EV, E. P., 1966, *Soviet Phys.*, *Solid St.*, **8**, 368.

RAJAGOPAL, A. K., and MAJUMDAR, C. K., 1970, *Prog. theor. Phys.*, **44**, 13.

RAMA REDDY, K., and CARRIGAN, R. A., 1970, *Nuovo Cim.*, **66**, 105.

REA, R. S., and DE REGGI, A. S., 1972, *Physics Lett.* A, **40**, 205.

RINDLER, W., 1960, *Special Relativity* (Oliver & Boyd).

ROAF, D. J., 1962, *Phil. Trans. R. Soc.*, **255**, 135.

ROCKMORE, D. M., and STEWART, A. T., 1967, *Positron Annihilation* (New York : Academic Press), p. 259.

ROELLIG, L. O., 1967, *Positron Annihilation* (New York : Academic Press), p. 127.

ROSE, K. L., and DE BENEDETTI, S., 1965, *Phys. Rev.*, **138**, A927.

ROUSE, L. J., and VARLASHKIN, P. G., 1971, *Phys. Rev.* B, **4**, 2377.

RUDGE, W. E., 1969, *Phys. Rev.*, **181**, 1024.

RUPAAL, A. S., and PATRICK, J. R., 1972, *Physics Lett.* A, **38**, 387.

SCHRADER, D. M., 1970, *Phys. Rev.* A, **1**, 1070.

SCHULMAN, J. H., and COMPTON, W. D., 1962, *Colour Centres in Solids* (Oxford : Pergamon Press).

SCHULTE, C. W., LICHTENBERGER, P. C., GINGERICH, R. R., and CAMPBELL, J. L., 1972, *Physics Lett.* A, **41**, 305.

SEDOV, V. L., TEIMURAZOVA, V. A., and BERNDT, K., 1970, *Physics Lett.* A, **33**, 319.

SEEGER, A., 1972 a, *Physics Lett.* A, **40**, 135 ; 1972 b, *Ibid.*, **41**, 267 ; 1973, *J. Phys.* F, **3**, 248.

SEITZ, F., 1954, *Rev. mod. Phys.*, **26**, 7.

SELIGER, H. H., 1955, *Phys. Rev.*, **100**, 1029.

SEN. P. and PATRO, A. P., 1969, *Nuovo. Cim.* B, **64**, 324.

SENICKI, E. M. D., BECKER, E. H., GOULD, A. G., and HOGG, B. G., 1972, *Physics Lett.* A, **41**, 293.

SENICKI, E. M. D., BECKER, E. H., GOULD, A. G., WEST, R. N., and HOGG, B. G., 1973, *J. Phys. Chem. Solids*, **34**, 673.

SHAND, J. B., 1969, *Physics Lett.* A, **30**, 478.

SINGH, K. P., SINGRU, R. M., TOMAR, M. S., and RAO, C. N. R., 1970, *Physics Lett.* A, **32**, 10.

SINGWI, K. S., TOSI, M. P., LAND, T. H., and SJÖLANDER, A., 1968, *Phys. Rev.*, **176**, 589.

SJÖLANDER, A., and STOTT, M. J., 1970, *Solid St. Commun.*, **8**, 1811 ; 1972, *Phys. Rev.* B, **5**, 2109.

SMITH, R. A., 1969, *Wave Mechanics of Crystalline Solids*, (London : Chapman & Hall).

SNEAD, C. L., GOLAND, A. N., KUSMISS, J. H., HUANG, H. C., and MEADE, R., 1971, *Phys. Rev.* B, **3**, 275.

SNEAD, C. L., HALL, T. M., and GOLAND, A. N., 1972, *Phys. Rev. Lett.*, **29**, 62.

SPEKTOR, D. M., and PAUL, D. A. L., 1971, *Proc. Second Int. Conf. on Positron Annihilation*, Kingston, Ontario, September, p. 5.71. (unpublished).

SPEKTOR, D. M., PAUL, D. A. L., and STEVENS, J. R., 1971, *Can. J. Phys.*, **49**, 939.

STACHOWIAK, H., 1970, *Phys. Stat. Sol.*, **41**, 599.

STANCANELLI, A., and FERRANTE, G., 1970, *Nuovo Cim.* B, **68**, 137.

STELDT, F. R., and VARLASHKIN, P. G., 1972, *Phys. Rev.* B, **5**, 4265.

STERN, E. A., 1968, *Phys. Rev.*, **168**, 730.

STEVENS, J. R., and MAO, A. C., 1970, *J. appl. Phys.*, **41**, 4273.

STEWART, A. T., 1957, *Can. J. Phys.*, **35**, 168 ; 1964, *Phys. Rev.* A, **133**, 1651 ; 1967, *Positron Annihilation*, (New York : Academic Press), p. 17.

STEWART, A. T., and BRISCOE, C. V., 1967, *Positron Annihilation* (New York : Academic Press), p. 387.
STEWART, A. T., and MARCH, R., 1961, *Phys. Rev.*, **122,** 75.
STEWART, A. T., and POPE, N. K., 1960, *Phys. Rev.*, **120,** 2033.
STEWART, A. T., and SHAND, J. B., 1966, *Phys. Rev. Lett.*, **16,** 261.
STEWART, A. T., SHAND, J. B., DONAGHY, J. J., and KUSMISS, J. H., 1962, *Phys. Rev.*. **128,** 118.
STROUD, D., and EHRENREICH, H., 1968, *Phys. Rev.*, **171,** 399.
SUEOKA, O., 1969, *J. phys. Soc. Japan*, **26,** 863.
SUEOKA, O., and ISHIHARA, H., 1972, *Physics Lett.* B, **42,** 131.
TAKHAR, P. S., 1967, *Phys. Rev.*, **157,** 257.
TANIGAWA, S., NANAO, S., KURIBAYASHI, K., and DOYAMA, M., 1971, *J. phys. Soc. Japan*, **31,** 1689 ; 1972, *Solid St. Commun.*, **10,** 1025.
TAO, S. J., 1970, *J. chem. Phys.*, **52,** 752 ; 1972, *Ibid.*, **56,** 5499.
TAO, S. J., and GREEN, J. H., 1964, *Proc. phys. Soc.*, **85,** 463.
THOMPSON, A., MURRAY, B. W., and BERKO, S., 1971, *Physics Lett.* A, **37,** 461.
THOMPSON, J. C., 1968, *Rev. mod. Phys.*, **40,** 704.
THOSAR, B. V., KULKARNI, V. G., LAGU, R. G., and CHANDRA, G., 1969, *Physics Lett.* A, **28,** 760.
TRIFTHÄUSER, W., 1971, *Proc. Second Int. Conf. on Positron Annihilation*, Kingston, Ontario, September, p. 4.77 (unpublished).
TUMOSA, C. S., NICHOLAS, J. B., and ACHE, H. J., 1971, *J. phys. Chem.*, **75,** 2030.
TURNBULL, D., and COHEN, M. H., 1961, *J. chem. Phys.*, **34,** 120.
UGGERHØJ, E., and ANDERSEN, J. U., 1968, *Can. J. Phys.*, **46,** 543.
VARLASHKIN, P. G., 1968, *J. chem. Phys.*, **49,** 3088.
VARLASHKIN, P. G., and ARIAS-LIMONTA, J. A., 1971, *J. chem. Phys.*, **54,**
VARLASHKIN, P. G., and STEWART, A. T., 1966, *Phys. Rev.*, **148,** 459.
VASHISHTA, P., and SINGWI, K. S., 1972, *Phys. Rev.* B, **6,** 875.
WAGNER, R. T., and HEREFORD, F. L., 1955, *Phys. Rev.*, **99,** 593.
WALKER, W. W., MERRITT, W. G., and COLE, G. D., 1972, *Physics Lett.* A, **40,** 157.
WALLACE, P. R., 1960, *Solid St. Phys.*, **10,** 1.
WATTS, B. R., 1964, *Proc. R. Soc.* A, **282,** 521.
WEISBERG, H., and BERKO, S., 1967, *Phys. Rev.*, **154,** 249.
WESELOWSKI, J., ROZENFELD, B., and SZUSKIEWICZ, M., 1963, *Acta phys. pol.*, **24,** 729.
WEST, R. N., 1971, *Solid St. Commun.*, **9,** 1417.
WEST, R. N., BORLAND, R. E., COOPER, J. R. A., and CUSACK, N. E., 1967, *Proc. phys.*, *Soc.*, **92,** 195.
WEST, R. N., CRISP, V. H. C., DE BLONDE, G., and HOGG, B. G., 1973, *Physics Lett.* (submitted).
WEST, R. N., and CUSACK, N. E., 1967, *Positron Annihilation* (New York : Academic Press), p. 309.
WHEELER, J. A., 1946, *Ann. N.Y. Acad. Sci.*, **48,** 219.
WICK, G., 1950, *Phys. Rev.*, **80,** 268.
WILLIAMS, D. Ll., BECKER, E. H., PETIJEVICH, P., and JONES, G., 1968, *Phys. Rev. Lett.*, **20,** 448.
WILLIAMS, R. W., LOUCKS, T. L., and MACKINTOSH, A. R., 1966, *Phys. Rev. Lett.*, **16,** 168.
WILLIAMS, R. W., and MACKINTOSH, A. R., 1968, *Phys. Rev.*, **168,** 679.
WILLIAMS, T. L., and ACHE, H. J., 1969, *J. chem. Phys.*, **51,** 3536.
WOLL, E. J., and CARBOTTE, J. P., 1967, *Phys. Rev.*, **164,** 985.
ZIMAN, J. M., 1961, *Adv. Phys.*, **10,** 1 ; 1965, *Principles of the Theory of Solids* (Cambridge University Press) ; 1967, *Adv. Phys.*, **13,** 89 ; 1969, *Elements of Advanced Quantum Theory* (Cambridge University Press).
ZUCHELLI, A. J., and HICKMAN, T. G., 1964, *Phys. Rev.* A, **136,** 1728.

Subject Index

Alkali halides—angular correlation, 26
  lifetimes, 77, 85
  defect studies, 80ff
Alkali metals—angular correlation, 41, 42, 53
  under pressure, 57
Alkali metal—ammonia solutions, 55, 56
Alloys, 56, 66, 67, 102, 103
Aluminium, 28, 47, 52, 67, 88, 90, 91, 94ff, 102
Angular correlation—anisotropy, 57
  deformed metals, 86, 87, 90
  diffraction effects, 12
  effect of pressure, 55
  ionic crystals, 53
  liquid metals, 103, 104
  metals, 52ff
  metals hydrides, 55
  molecular materials, 53
  polyatomic systems, 49
  semiconductors, 52ff
  temperature dependence metals, 86, 87, 90, 100, 110, 111
  theory, 5ff, 27, 28, 46ff
Angular correlation techniques—
  long slit geometry, 10ff
  multi detector systems, 61
  point detector geometry, 60, 61
  short slit geometry, 59, 60
Annihilation centres, 26, 27, 71
  —see also Trapping
Annihilation in matter, theory, 46ff
Annihilation rate—see Lifetime(s)
Annihilation—with conduction electrons, 27ff
  with tightly bound electrons—
  see Core annihilation
Aqueous solutions, 23
Argon, 18, 48

Beryllium, 28, 58, 69
Bismuth, 28, 57, 103

Cadmium, 28, 87ff, 94, 99
Cerium, under pressure, 55
Cerium hydride, 55
Cobalt, 28
Compton scattering, 51, 65
Condensed gases, 17
Conduction electron annihilation, 27ff
Conversion—see Quenching
Copper—angular correlation, 47, 48, 58ff
  Fermi surface, 59ff
  lifetime, 28
  alloys, 66, 67

Core annihilation, 28, 47, 48, 54, 56, 65
Cross sections for annihilation—
  one photon, 2
  two photon, 2, 3
  three photon, 2, 4
Cyclohexane, phase transition, 25

Defect sensitive annihilation parameters, 75
Diamond, lifetime, 40
Doppler broadening, 6, 12, 23, 57, 61, 75, 101, 111

Effective mass of positrons in metals, 40ff
Electric field effect, 15
Electron – positron interaction, 29ff
Experimental techniques—
  angular correlation, 10ff, 59ff
  Doppler broadening, 12
  lifetime, 8ff
  three photon rate, 16

Fermi ' cut off ', 28
Fermi surface studies, 57ff
Ferromagnetic materials, 70, 71
Free electron approximation, 27
Free volume model, 20, 21

Gadolinium, 55, 71
Gallium, 28, 103
Germanium, 28, 40, 57
Gold, 93
Graphite, 57
Green's functions, 30

Helium, liquid—angular correlation, 19
  lifetime, 18, 19
Higher momentum components—
  electron wavefunctions, 50, 51, 60
  positron wavefunctions, 50
  positronium wavefunctions, 80
Holmium, 70
Hydrocarbons, 24
Hydrides, 26, 49, 55

Ice, 24, 76
Indium, 28, 86, 88, 91, 94, 96, 99
Ionic crystals, 25, 27, 76ff
  —lifetimes, 77
Iron, 28, 71, 89
Irradiation damage, 80ff, 89, 101

Lead, 28, 99
Lifetime(s)—effect of lattice vibrations, 21, 100
  ionic crystals, table of, 77

liquid metals, 104, 105
metals—theory, 27ff
    table of, 28
semiconductors, 40
temperature dependence in
    metals, 87, 88, 100
Lifetime spectra—analysis, 10, 74, 98ff
        defected metals, 91
        ionic solids, 13, 85
        molecular materials, 15, 16,
            24
        powders, 105
        theoretical forms, 71ff
Liquid metals, 103ff
Lithium, 28, 51, 68
Lithium hydride, angular correlation, 26, 49

Magnesium, 28, 57
Magnetic field effects—quenching, 16, 79
        focusing, 11
        ferromagnetic
            materials, 70, 71
Many electron effects—angular correlation,
            33, 34, 38ff
        lifetime, 29ff
        positron behaviour,
            40ff
Mercury—angular correlation, 103
        lifetime, 28, 104, 105
Molecular liquids, 18, 19
Molecular solids, 19ff
Molybdenum, 28, 70, 101, 108

Naphthalene, 23
Nickel, 28, 71, 101
Nitrogen, liquid, 17
Noble metals, angular correlation, 53

Ore gap, 14, 25
Organic liquids, 18, 19
Orthopositronium—see Positronium
Oxides, 77, 105

Palladium—hydrogen, 55
Parapositronium—see Positronium
Parapositronium—peaks in angular distribu-
            tion, 16, 17, 106ff
        satellite peaks, 78ff
Parity conservation, 3
Phase transitions, 23ff
Pick-off—see Quenching
Platinum, 89
Polarized positrons, 70
Polyethylene, 20, 22
Polymers, 19, 25
Positron—absorption, 109
        affinity, 25—see also Positron work
            functions
        back scattering, 109
        bound states, 26, 44, 76, 86
        channelling, 110

diffusion, 84, 86, 93, 106
effective mass in metals, 40ff
energy loss—see Thermalization
lifetimes—see Lifetimes
mobility, 92, 93
self energy, 35
sources, 10, 11, 14, 70, 105, 109, 110
thermalization—see Thermaliza-
    tion
trapping—see Trapping
wavefunctions, 44ff
work functions, 25, 108ff
Positronium—affinity, 27—see also
        Positronium work functions
        binding energy, 14, 25, 26
        bubbles, 18, 19, 24
        chemistry, 18
        formation—see Ore gap
        inhibition, 15
        in ionic crystals, 25, 76ff
        in metals, 27
        negative ion, 78
        ortho-lifetime, 4
        para-lifetime, 4
        quenching—see Quenching
        wavefunctions, 4
        work functions, 109, 110
Potassium, 28
Potassium chloride, 80ff
Powders, 105ff
Pressure effects, 55, 57

Quartz, 23, 78, 79, 106
Quenching—chemical, 15
        conversion, 15
        magnetic field, 16, 79
        pick-off, 15, 16

Rare earth metals, 55, 58
Rigid band model, 58
Rubidium, 28, 42

Selenium, 57
Silica gels, 107
Silicon, 28, 40, 57, 68
Silicon dioxide powders, 106, 107
Sodium, 28, 34, 41, 42
Spectrum analysis, 10, 74, 98ff
Spin density studies, 70, 71
Sulphur, 23

Thermalization, 14, 30, 40
Three photon annihilation, 2, 4, 16
Tin, 28, 57
Titanium—hydrogen, 55
Trapping by—density fluctuations, 72, 105
        dislocations, 87, 88, 97, 101
        F-centres, 81ff
        grain boundaries, 91
        impurities, 93, 89, 102, 103
        surfaces, 107ff

vacancies, 83, 84, 87ff, 93ff
voids, 71, 101, 106ff
Trapping in—alkali halides, 80ff
alkali metals, 90, 104
alloys, 102
ionic crystals, 27, 76ff
liquid metals, 103ff
metals—aluminium, 88, 90, 91, 94ff, 102
brass, 101
cadmium, 89, 94, 99
copper, 89
gold, 93
indium, 88, 91, 94, 99
iron, 89
lead, 99
molybdenum, 101, 108
nickel, 101
platinum, 89

tungsten, 90
zinc, 88, 94, 99
powders, 106ff
Trapping mechanisms, 91ff
Trapping models, 73, 74, 90, 91
Trapping rates, 72, 92
temperature dependence of, 92, 93, 98, 101
Two-photon annihilation, 5ff

Vacancy formation energy, 88, 96ff
V-centres, 80

Water, 23, 24

Yttrium, 69

Zinc, 28, 48